Disinfectant Decay and Corrosion: Laboratory and Field Studies

The mission of the Awwa Research Foundation (AwwaRF) is to advance the science of water to improve the quality of life. Funded primarily through annual subscription payments from over 1,000 utilities, consulting firms, and manufacturers in North America and abroad, AwwaRF sponsors research on all aspects of drinking water, including supply and resources, treatment, monitoring and analysis, distribution, management, and health effects.

From its headquarters in Denver, Colorado, the AwwaRF staff directs and supports the efforts of over 500 volunteers, who are the heart of the research program. These volunteers, serving on various boards and committees, use their expertise to select and monitor research studies to benefit the entire drinking water community.

Research findings are disseminated through a number of technology transfer activities, including research reports, conferences, videotape summaries, and periodicals.

Disinfectant Decay and Corrosion: Laboratory and Field Studies

Prepared by:
Francis A. DiGiano, Vanessa Speight, and **Weidong Zhang**
Department of Environmental Sciences and Engineering
University of North Carolina
Chapel Hill, NC 27599-7400

Sponsored by:
Awwa Research Foundation
6666 West Quincy Avenue
Denver, CO 80235-3098

Published by the

Awwa Research Foundation
Advancing the Science of Water®

IWA Publishing

DISCLAIMER

This study was funded by the Awwa Research Foundation (AwwaRF). AwwaRF assumes no responsibility for the content of the research study reported in this publication or for the opinions or statements of fact expressed in the report. The mention of trade names for commercial products does not represent or imply the approval or endorsement of AwwaRF. This report is presented solely for informational purposes.

Published by IWA Publishing, Alliance House, 12 Caxton Street, London SW1H 0QS, UK

Telephone: +44 (0) 20 7654 5500; Fax: +44 (0) 20 7654 5555; Email: publications@iwap.co.uk;
Web: **www.iwapublishing.com**

AwwaRF report number 90992F. Originally published by AwwaRF for its subscribers in 2003.
IWA Publishing version published 2004.

ISBN 1 84339 8877

CONTENTS

TABLES

FIGURES

FOREWORD

The Awwa Research Foundation is a nonprofit corporation that is dedicated to the implementation of a research effort to help utilities respond to regulatory requirements and traditional high-priority concerns of the industry. The research agenda is developed through a process of consultation with subscribers and drinking water professionals. Under the umbrella of a Strategic Research Plan, the Research Advisory Council prioritizes the suggested project based upon current and future needs, applicability, and past work; the recommendations are forwarded to the Board of Trustees for final selection. The foundation also sponsors research projects through an unsolicited proposal process; the Collaborative Research, Research Applications, and Tailored Collaboration programs; and various joint research efforts with organizations such as the U.S. Environmental Protection Agency, the U.S. Bureau of Reclamation, and the Association of California Water Agencies.

This publication is a result of one of these sponsored studies, and it is hoped that its findings will be applied in communities throughout the world. The following report serves not only as a means of communicating the results of the water industry's centralized research program but also as a tool to enlist the further support of the nonmember utilities and individuals.

Projects are managed closely from their inception to the final report by the foundation's staff and large cadre of volunteers who willingly contribute their time and expertise. The foundation serves a planning and management function and awards contracts to other institutions such as water utilities, universities, and engineering firms. The funding for this research effort comes primarily from the Subscription Program, through which water utilities subscribe to the research program and make an annual payment proportionate to the volume of water they deliver and consultants and manufacturers subscribe based on their annual billings. The program offers a cost-effective and fair method for funding research in the public interest.

A broad spectrum of water supply issues is addressed by the foundation's research agenda: resources, treatment and operations, distribution and storage, water quality and analysis, toxicology, economics, and management. The ultimate purpose of the coordinated effort is to assist water suppliers to provide the highest possible quality of water economically and reliably. The true benefits are realized when the results are implemented at the utility level. The foundation's trustees are pleased to offer this publication as a contribution toward that end.

Water utility interest in applying membrane technologies continues to grow. Recent advances in membrane technology, regulatory pressure, decreased costs, and the growing number of facilities applying membrane technologies are driving this interest. One potential limitation to applying these technologies is the tendency for membranes to foul. Fouling leads to decreased permeate water flux or an increase in applied pressure to maintain the flux. The costs of cleaning membranes, applying pressure, and backflushing the membranes can make the price of membrane technology prohibitive for water utilities. The Foundation undertook this project to investigate new membrane materials with less fouling tendency with the hope that in the future membranes would be more affordable for utilities to apply.

Edmund G. Archuleta, P.E.
Chair, Board of Trustees
Awwa Research Foundation

James F. Manwaring, P.E.
Executive Director
Awwa Research Foundation

ACKNOWLEDGMENTS

The authors of this report wish thank Randall Goodman, Instrument Shop Manager in the Department of Environmental Sciences and Engineering at the University of North Carolina who constructed the laboratory pipe-section reactor. The staff at the North Chatham County Water Treatment Plant in Pittsboro, N.C., the Department of Public Works in Durham, N.C. and the Department of Environmental Resources in Durham, N.C. are acknowledged for providing essential information on distribution system characteristics and for their assistance in sample collection. Finally, Pitometer Associates, Inc. is acknowledged for providing additional financial support to perform a water quality calibration of the EPANET model for the City of Durham.

EXECUTIVE SUMMARY

The deterioration of water quality in the large and complex reaction environment of the distribution system (DS) is cause for serious concern. Loss of disinfectant residual occurs through many different pathways that are not yet fully understood and this can lead to bacterial regrowth as well as other problems such as taste and odors. The disinfectant can be chemically reduced in the bulk water and at the walls of the pipe.

The disinfectant decay rate in the bulk water can be measured directly by placing the test water in a non-reactive container and measuring the loss in disinfectant residual with time. This is often described as a pseudo-first order reaction with respect to disinfectant residual concentration even though the rate process is very likely to depend upon the concentration of the reducing agent as well as the disinfectant. Far less is known about the surface reaction rate at the pipe wall and its dependency on such factors as the disinfectant residual concentration, pipe material, corrosion potential, and water velocity. Depending on the age and condition of the DS, wall reactions may account for most of the loss of disinfectant residual.

Despite the importance of wall reactions, an experimental protocol to measure the decay rate of disinfectant has yet to be standardized and widely adopted in the water utility industry. Indirect techniques have been proposed wherein the disinfectant residual is measured in the bulk water at two locations, A and B, on the same pipe line in the DS. A batch rate test is also performed on a grab sample to calculate the decay rate in the bulk water. A wall reaction rate expression is assumed (e.g., a zero- or first-order rate with respect to disinfectant concentration) and a wall reaction rate coefficient is fit to explain the additional decrease in bulk disinfectant residual from location A to location B not accounted for by the bulk decay rate. This is often referred to as the "in-situ method." Several drawbacks to this method are important to note. Foremost, this test requires a sufficiently long pipe in the DS to see a measurable drop in disinfectant residual, with few connections to minimize mixing effects; these locations can be difficult to find. Moreover, focusing on individual pipe sections makes it difficult to quantify a general relationship of decay coefficients with water quality, corrosion or water velocity.

The other common alternative is to measure the disinfectant residual at many locations throughout the distribution system and then use these data together with a hydraulically calibrated network model to calibrate a disinfectant residual sub-model by finding the wall reaction coefficients that best describe the disinfection residual data. Modeling software such as EPANET has incorporated an option which calculates the wall reaction coefficient based on an inversely proportional multiplier for the Hazen-Williams C-factor for each pipe segment. Alternatively, the user of such software can assign wall reaction coefficients to groupings of pipes (e.g., depending on material and age) or to individual pipe segments.

While the decay of free chlorine in distribution systems has been widely reported, far less is known about the decay of chloramines. Given the increasing popularity of chloramine disinfection to control disinfectant by-product formation, the water utility industry needs to have a better understanding of chloramine stability in the distribution system. The available data indicates that chloramines residual is often lost by chemical reduction at the pipe surface, especially where corrosion is active. However, much more data for all type of pipe materials are needed, as well as a better understanding of the underlying mechanisms, to fully characterize the chloramine decay reactions.

The literature includes both first-order and zero-order rate models with respect to disinfectant residual to describe wall reactions. Different models are thought to be required to describe varying conditions in the distribution system. The theory that corrosion products serve as reducing agents for disinfectant and may govern the decay rate under certain conditions leads to a zero-order rate description with respect to disinfectant residual. Reactions where disinfectant residual concentrations govern the reaction rate can be better described by a first-order reaction. Therefore, a better understanding of the relationship between corrosion and disinfectant decay rate is particularly important for proper selection of the rate model to be used in such network water quality models as EPANET.

Once a rate model has been selected for use in a network water quality model, the user must then define reaction rate coefficients. A few rate coefficients have been published in the literature for selected distribution system studies. However, the distribution systems studied represent only a small number of water quality conditions. The database of rate coefficients needs to be expanded to facilitate widespread use of water quality models.

RESEARCH OBJECTIVES

The primary objective of this research was to develop and test a pipe-section reactor to measure the decay rate of disinfectant at the pipe wall. The pipe-section reactor is more convenient than pipe loop reactors which require a large laboratory and large supply volume of water. It is also a more realistic representation of the pipe surface than provided in pipe coupon experiments or annular reactors. The effects of water velocity and corrosion rate can be investigated using a pipe-section reactor, while still affording the simplicity of laboratory-scale, batch reactor operation. The secondary objective was to collect field data from two distribution systems in North Carolina (North Chatham County and Durham) to measure the decay of free chlorine by two alternative methods and to explore a relationship with corrosion rate.

The ultimate goal would be to achieve a better understanding of disinfectant decay at the pipe walls such that a set of default values for decay rate coefficients could be specified for free chlorination and chloramination, the two most common secondary disinfectants. These coefficients could be available within modeling software to be assigned to pipe sections based on information about pipe materials, corrosion rates, water velocities and key water quality measures (e.g., pH, alkalinity and dissolved oxygen). This would ensure that the wall reaction has been properly accounted for within the model. Ultimately, an accelerated model calibration process and greater confidence in model predictions would be valuable for water utilities when evaluating the impact of different strategies such as pipe replacement, alternative storage tank management and changes in water treatment technology to meet both disinfection and disinfection by-product regulations.

APPROACH

The pipe-section reactor consisted of a 20-inch (508 mm) length of either a six-inch (152 mm) diameter section of cast iron (CI) or cement-lined ductile iron (DI) pipe. An inner cylinder of Plexiglas was inserted to create a narrow annular space between pipe wall and the Plexiglas wall. A large, propeller-type stirrer was inserted into the inner cylinder to circulate water from the inner cylinder to the annular space. The water velocity adjacent to the pipe wall was controlled by the rotational speed of the stirrer motor; the velocity range was from about 0.5 to

2.0 ft/s (0.16 to 0.62 m/s). The pipe-section reactor can be operated in both continuous flow and batch modes of operation. In either mode, experiments can be designed easily to test different rate models for wall reactions.

The reactor was filled with the tap water distributed by the Orange Water and Sewer Authority (OWASA) which on annual average is of very low alkalinity (3.1 mg/L as $CaCO_3$) with a pH of 7.4 and a total organic carbon (TOC) of 3.2 mg/L. The average annual concentration of a phosphate-based corrosion inhibitor was 1.3 mg/L. Except in experiments in which dissolved oxygen (DO) was removed, the DO was between 7 and 8.5 mg/L in the test water within the reactor at the beginning of the experiment. The decay rate in bulk water was measured in independent experiments in which 2 L of OWASA tap water was placed in a 2-L clean glass bottle in a dark place at room temperature (21°C). Sample volumes of 25 mL were withdrawn for each chlorine measurement, giving a total withdrawal of 200 to 300 mL in each rate test.

The experiments with the CI pipe-section reactor measured the effects of initial chlorine concentration, water velocity, pH, Cl^- and DO on the rate of chlorine decay. The same parameters with the exception of DO were investigated in a DI pipe-section reactor.

With regard to the field experiments, both distribution systems were supplied by surface waters that had been treated by alum coagulation, sedimentation and filtration. The finished waters were of low alkalinity and hardness (<30 mg/L as $CaCO_3$ of either parameter) and moderate TOC (2 to 5 mg/L). Both utilities add a phosphate-based corrosion inhibitor.

The North Chatham County system serves about 2500 customers. The DS includes a long section of DI pipe with no significant water demands so that free-chlorine wall demand could be estimated by the in-situ method; corrosion rate was also measured. Two measurements were made in October and November of 2000.

The Durham system is much larger than the North Chatham County system, serving about 150,000 customers. Because of the difficulty in isolating a single long run of pipe with few connections, the alternative approach was to use a hydraulically calibrated, 12,000 pipe EPANET model and to calculate chlorine decay coefficients for the wall reaction using results from approximately 100 measurements of free chlorine residual from the DS in July 2001.

CONCLUSIONS

1. The pipe-section reactor hold promise for quantifying a governing relationship among pipe characteristics (e.g., C-factor, age, material, corrosivity), water quality, water velocity and $k_{w.}$
2. A zero-order kinetic model described the overall chlorine decay rate for CI pipe quite well for a range of initial chlorine concentrations, pH, Cl^- concentration and DO concentration. A zero-order model would be consistent with wall reactions that dominate the total chlorine decay rate and rate control by the generation of corrosion products rather than the concentration of chlorine.
3. Chlorine decay tests in glass bottles confirmed that the decay rate in bulk water was much slower than the overall decay in the pipe-section reactor and was described by first-order kinetics with respect to chlorine concentration.
4. Factors that increased the zero-order rate coefficient for decay at the pipe wall were higher velocity (due to higher mass transfer at the pipe surface), lower pH (due to faster corrosion rate), and lower DO (due to greater release of Fe^{2+} after removal of the oxidized layer at the surface). Although increasing the Cl^- concentration greatly

increased the corrosion rate, it did not produce an immediate in increase the zero-order rate coefficient. This observation points to the difficulty of using corrosion rate measurements in bulk water as a predictor of wall reactions.

5. Unlike the CI pipe section, the chlorine decay kinetics were first-order with respect to chlorine concentration in the DI pipe section. While the wall reactions were still important contributors to chlorine decay, the rate of decay was about five times slower in the DI pipe section than in the CI pipe section. The shift to first-order decay kinetics could suggest that chlorine becomes the limiting reactant in new, cement-lined DI pipe.
6. Field studies of the two distribution systems were difficult to compare directly because different techniques (in-situ method vs. EPANET calibration method) had to be used to investigate the relationship between corrosion rate and disinfectant decay in each. Corrosion rates as measured in bulk water samples using the AquaMate Portable Corrater®, ranged from 4.3 to 6.1 mpy. These would be considered low to moderate values based on other field studies. Nevertheless, insights were gained not only into the importance of corrosion reactions in reducing disinfectant residual but also into the advantages and disadvantages of the two measurement methods to characterize the reactions.
7. Electrochemical measurements of corrosion rate with bulk water samples do not account for on-going surface reactions at pipe walls and thus, such information is of limited value.
8. The corrosion rates in the North Chatham County system on two sampling dates were 4.3 mpy and 5.1 mpy. The lower corrosion rate corresponded to a temperature that was 7°C lower. The calculated first-order chlorine decay coefficients for wall reaction, k_w, were 0.094 ft/d and 0.154 ft/d for the corrosion rates of 4.3 mpy and 5.1 mpy, respectively. Temperature played a role in both wall reaction and corrosion rate in this case.
9. In the Durham system, use of an EPANET model to determine k_w gave a wide range of values even within one type of pipe material (0.25 to 1.23 ft/d for CI pipe and from 0.17 to 2 ft/d for DI, comprising both and old new pipe). Little further interpretation is possible for the range in k_w owing to the uncertainty associated with the calibration method and the fact that fewer than 100 locations scattered throughout the DS were used for free chlorine residual measurements. A greater number of samples would lessen the uncertainty is estimates of k_w for each pipe segment.
10. The k_w values for old DI pipe for the Durham and North Chatham County distribution systems were similar (0.21 – 0.25 ft/d vs. 0.285 – 0.463 ft/d) although measured by two different methods (in-situ method in North Chatham County vs. EPANET calibration in Durham). The corrosion rates were somewhat different (4.3 mpy – 5.1 mpy in North Chatham County vs. 6.1 - 6.4 mpy in Durham). The data are too limited to draw further conclusions; however, a relationship between corrosion and free chlorine decay is possible.
11. The field studies did not provide a specific relationship that would form the basis of a method for estimating k_w from measurement of corrosion rate in the bulk water. The results showed the importance of unique characteristics found in every DS that dictate the approach to investigating k_w and the inherent drawbacks to corrosion rate measurements.

RECOMMENDATIONS

1. Measurements of wall demand rates determined in distribution systems using the *in-situ* methodology and in the laboratory using the pipe-section reactor should be compared to demonstrate the utility of the latter method.
2. Beyond a comparison study of two methods, field studies that are intended to increase the database on k_w and its relationship to corrosion would be costly and time-consuming and they would probably still not be of general use to all water utilities. Despite the costs, field studies will nevertheless be of value in specific cases.
3. The pipe-section reactor studies conducted in this research should be greatly expanded to examine a wide range of pipe materials and age, water velocity and disinfectant type (e.g., chloramines vs. free chlorine).

CHAPTER 1
INTRODUCTION AND OBJECTIVES

The deterioration of water quality in the large and complex reaction environment of the distribution system is cause for serious concern. Loss of disinfectant residual occurs through many different pathways that are not yet fully understood and this leads to bacterial regrowth as well as other problems such as taste and odors. Unless this problem is addressed, water customers may not benefit from large investments to reduce health risks through new water treatment technology. Pipe surfaces are largely responsible for the complexity of the reaction environment. These surfaces are frequently coated with a variety of scales whose composition depends on pipe type, source water quality (especially ground vs. surface), and treatment techniques. Scales begin with reactions that occur on surface of bare pipe. Iron and steel pipes produce iron corrosion scales on the inner surface.

The loss of disinfectant residual within distribution systems can be dominated by reactions on corrosion scales where significant corrosion is present. This source of disinfectant loss is commonly referred to as *wall reaction* to differentiate it from *bulk demand*. A pseudo-first order reaction with respect to disinfectant concentration is commonly used to describe the bulk reaction rate although the exact nature of the reaction is still not well understood. Less is known about the wall reaction rate. Myriad conditions control wall reaction rates. These may include the chemical composition of the corrosion products, water velocity, temperature and chemical characteristics of the bulk water.

Water quality computer models of the distribution system can be used to calculate rate coefficients for wall reaction. Once the hydraulic portion of the model has been calibrated and a rate coefficient for bulk water demand is measured independently in laboratory experiments, the rate coefficient for wall demand becomes the calibration parameter within the model. While success has been reported in several case histories, the technique lacks a firm foundation in a kinetic formulation for wall reactions and requires extensive field tests and repeated computer runs to achieve calibration.

The alternative to wall demand calibration with water quality models is a systematic evaluation of wall demand in laboratory tests where causative factors can be isolated for study. Because the reaction occurs on a surface, the reaction rate expression needs to account for the available surface area, which can be difficult to quantify when corrosion scales form large and uneven tubercles. For the same reason, use of small coupons of pipe material will not simulate the complexity of the actual reactive surface. Proper simulation of water velocity is also needed in laboratory tests because mass transfer of reactant (disinfectant) from the bulk water through a *hydrodynamic boundary layer* to the pipe surface is important. The thickness of the boundary layer in turn depends upon water velocity. In addition, the rate of appearance of corrosion products at the pipe surface will depend upon the corrosion rate that is influenced by many factors related to water chemistry, pipe material and pipe age.

Although a few references exist in the literature, the site-specific nature and variability of wall demand coefficients make it difficult for users to find and apply values for their own modeling applications. Ideally, it would be useful to have a library of rate coefficients available within the modeling software which could be assigned to pipe sections based on information about pipe materials, corrosion rates, water velocities and key water quality measures (e.g., pH, alkalinity and dissolved oxygen). The use of these reaction rate coefficients would assure that the wall demand has been properly accounted for within the model, accelerate the calibration process, and lead to more confidence in model predictions.

Although the reactions that cause chlorine decay are still not well understood, even less is known about reactions that cause chloramine decay. Given the trend toward replacement of free chlorine with chloramine as the secondary disinfectant in order minimize the formation of disinfection by-products, water utilities need to have a better understanding of chloramine stability in the distribution system. An important part of the research needed on chloramine stability is an assessment of chloramine wall reactions.

The primary objective of this research was development and testing of a pipe-section reactor to measure wall demand that is simpler than typical pipe loop experiments and more realistic than pipe coupon experiments. The reactor design enables simulation of water velocity and measurement of corrosion rate while still affording the simplicity of laboratory-scale, batch reactor operation. The secondary objective was to collect field data from two distribution systems in North Carolina to explore further the nature of wall demand exerted by both free chlorine and chloramines. The North Chatham County system (2,500 customers) provided a long section of ductile iron pipe with no significant water demands in order to use what is sometimes referred to as the in-situ method to calculate free-chlorine wall demand from measurements of total chlorine loss; corrosion rate was also measured. Wall demand from the North Chatham County system was compared to decay in the City of Durham (150,000 customers). Because of the difficulty in isolating a single pipe with few service connections, the in-situ method was not used to calculate wall demand coefficients in Durham. The alternative approach was to use a hydraulically calibrated, 12,000 pipe EPANET model for calibration of free chlorine decay based on measurements of free-chlorine residuals throughout the distribution system.

CHAPTER 2
LITERATURE REVIEW

THE NATURE OF CORROSION

Corrosion is a reaction that concerns the dissolution of iron from pipe surfaces:

$$Fe \rightarrow Fe^{2+} + 2e^- \quad (2.1)$$

For this process to occur, electron acceptors are needed. The mostly commonly considered electron acceptor are dissolved oxygen and hydrogen ions:

$$O_2 + 4e^- + 2H_2O \rightarrow 4OH^- \quad (2.2)$$

$$2H^+ + 2e^- \rightarrow H_2 \quad (2.3)$$

Once areas of the surface become oxidized, the surface acts an electrochemical cell with anodic (attracting negatively charged ions) and cathodic (attracting positively charge ions) regions. The aqueous phase adjacent to the pipe is the electrolyte that allows transport of electrons from the anodic to cathodic regions. The subsequent reactions of Fe^{2+} with several different chemical species in the water lead to scale layers or deposits. Those closest to the pipe wall are reduced while those farthest away are oxidized. Some typical deposits that contain Fe in a reduced state near the pipe wall are siderite ($FeCO_3$), pyrhotite (FeS) and green rust, a Fe^{2+}, Fe^{3+} hydroxide. Goethite (α-FeOOH), lepidocrocite (γ-FeOOH) and magnetite or black rust (Fe_3O_4) would be found in the outermost, oxidized deposits. Dissolved oxygen, alkalinity, temperature, pH, dissolved inorganic ions, organic matter, microbial activity as well as other water quality characteristics affect the rate of corrosion and the chemistry of deposits (Snoeyink and Wagner 1996).

Sarin et al. (2001) describe the physical and chemical characteristics of corrosion scales on 90-year old cast iron pipe and 70-year-old galvanized pipe sections removed from two different distribution systems. In both samples, the corrosion scales comprised a dense shell-like layer that covered a porous interior of the scales, separating the porous region from the bulk water. Table 2.1 summarizes the major physical and chemical characteristics of these two materials. Iron is the dominant element in the scales. The porosity, pore size range and specific surface area are all indicators of the potential for transport of reactants and products in these scales. However, mass transfer could limit reactions within the porous interior of the scales. Wetting of scales shifted the form of Fe in the interior from Fe^{3+} to Fe^{2+} although the surface of the tubercles remained in the Fe^{3+} form. The presence of ferrous hydroxide [$Fe(OH)_2$] between the shell-like layer and the pipe wall was postulated. Oxygen in water could promote the oxidation of this Fe^{2+} phase to a Fe^{3+} phase such as magnetite at the shell-water interface.

In related work, Sarin et al. (2000) showed that under anaerobic conditions, Fe^{3+} in the shell-like layer was reduced to Fe^{2+} which weakened the shell and allowed large quantities of Fe^{2+} from deeper layers to be released into the water. Their experiments did not include reactions with free chlorine or chloramines. The reducing conditions that cause release of Fe^{2+} from

Table 2.1
Physical and chemical characteristics of corrosion scales

	Galvanized steel pipe*	Cast iron pipe†
Bulk density (g/cm^3)	2.2–2.3	1.7–2.2
Porosity (%)	40–43	44–54
BET surface area (m^2/g)‡	89–154	32–54
Average pore diameter (Å)	32–39	129–160
Elemental composition (dry form testing)	Fe predominates	Fe predominates
Top layer mineral forms (dry form testing)	magnetite, goethite	magnetite, geothite, lepidocrocite

Data from: Sarin et al. 2001.
* From Northern Illinois Water Corporation distribution system in Champaign, Ill.
† From Metropolitan Water Resources Authority distribution system in Boston, Mass.
‡ Brauner, Emmet and Teller specific surface area

beneath the protective shell-like layer of corrosion scale, such as evidenced by low oxygen concentrations, should increase the disinfectant decay rate significantly. While the rate of iron release is important with respect to promotion of disinfectant decay, it not necessarily linked to the rate of corrosion (Sarin et al. 2000). Iron release includes not only the Fe^{2+} from the pipe surface, which is related to the corrosion rate, but also Fe^{2+} from within the corrosion scale.

Frateur et al. (1999) photographed the corrosion scales formed in the presence and absence of free chlorine. They observed that when free chlorine was not present, a reddish layer covered the metallic surface whereas this layer was far less evident in the presence of free chlorine. The reddish layer is hematite (α-Fe_2O_3) and goethite (α-FeOOH), the latter originating from oxidation of Fe^{2+} by oxygen. Below the surface was a porous layer of black rust (Fe_3O_4) and some green rust (mixture of Fe^{2+} and Fe^{3+}). These observations are similar to those of Sarin et al. (2001). Frateur et al. propose that the reaction of free chlorine with Fe^{2+} limits the formation of red rust because it is much faster than the reaction of oxygen with Fe^{2+} to form geothite.

REACTIVITY OF FREE CHLORINE AND CHLORAMINES WITH CORROSION PRODUCTS

The stoichiometry for oxidation of Fe^{2+} by free chlorine is:

$$2Fe^{2+} + HOCl \rightarrow 2Fe^{3+} + Cl^- + H_2O \qquad (2.4)$$

An empirical rate law has been proposed for bulk water reaction between Fe^{2+} and free chlorine (Knocke et al. 1990):

$$-\frac{d[Fe^{2+}]}{dt} = k_{HOCl}[Fe^{2+}]^{0.96}[HOCl]^{1.65}[OH^-]^{0.32} \qquad (2.5)$$

where $k_{HOCl} = 8.4 \times 10^{16}\ M^{-1.93}min^{-1}$

The units for k_{HOCl} (1.93 L/mole/min) are needed to express the rate of change of Fe^{2+} correctly as mole/L/min from the orders of reaction with respect to Fe^{2+}, HOCl and OH^-. As seen by the magnitude of $k_{HOCl,}$ this reaction is extremely fast.

In contrast to Equation 2.5, Kiene and Wable in an AwwaRF project (Vasconcelos et al. 1996) proposed a zero-order rate for free chlorine decay that was determined by the corrosion rate. They assumed that free chlorine was the only oxidant consumed in the corrosion reaction and that the reaction is limited by the generation rate of Fe^{2+} (i.e., free chlorine reacts very fast with Fe^{2+} once it appears at the pipe surface). Mass transfer of free chlorine to the pipe wall was not considered rate limiting given that the corrosion reaction has a half life measured on the order of hours whereas the time for molecular diffusion of free chlorine is on the order of seconds. Assuming that the corrosion rate is directly proportional to the rate of Fe^{2+} production gives:

$$-\frac{d[\mathrm{Cl}]}{dt} = \frac{4}{D} \times \frac{M_{Cl}}{M_{Fe}} \times \rho_{Fe} \times V_{cor} \tag{2.6}$$

where

D = pipe diameter (m)
$4/D$ = surface area-to-volume ratio
M_{Cl} = atomic weight of chlorine (35.5 g/mol)
M_{Fe} = atomic weight of iron (55.85 g/mol)
ρ_{Fe} = density of iron (7860 kg/m^3)
V_{cor} = corrosion velocity (µm/year)

Expressing the free chlorine decay rate in mg/L/sec gives:

$$-\frac{d[Cl]}{dt} = \frac{4 \times 10^{-3}}{\tau} \times \frac{M_{Cl}}{M_{Fe}} \times \rho_{Fe} \times \frac{V_{cor}}{D} \tag{2.7}$$

where 10^{-3} is to convert from m^3 to L and τ is to convert from seconds to year (3.16×10^7 sec/year). The sense of this simple rate expression is that $\rho_{Fe}V_{cor}/D$ is the loss rate of Fe from the pipe and M_{Cl}/M_{Fe} accounts for the stoichiometry between free chlorine and iron in Equation 2.1. Upon substitution of the known coefficients, the final rate expression is:

$$-\frac{d[Cl]}{dt} = 6.32 \times 10^{-7} \times \frac{V_{cor}}{D} \tag{2.8}$$

This same development can also be found in Kiene, Lu, and Levi (1998) and the same resulting equation can be derived from equations provided by Frateur et al. (1999).

The electrochemical consumption of free chlorine at the metal surface is often overlooked (Frateur et al. 1999). Free chlorine can serve as the electron acceptor for oxidation of elemental Fe to Fe^{2+}:

$$\mathrm{HOCl} + \mathrm{H}^+ \; 2e^- \rightarrow \mathrm{Cl}^- + \mathrm{H_2O} \tag{2.9}$$

While free chlorine is a more favorable acceptor than oxygen, this reaction is apparently considered as a small contributor to the overall decay rate of free chlorine (Frateur et al. 1999).

The Corrater SCA1 is an instrument that measures the polarization resistance which is then converted to a corrosion rate. Raad et al. (1998) noted, however, that the low conductivity (high resistivity) and relatively low corrosion rate of drinking water compared to industrial waters limits the accuracy of the linear polarization method. Nevertheless, they obtained strong correlations for a particular choice of electrode material with standard weight loss measurements and they were able to use the method to select the best corrosion control strategy.

Field experiments were conducted with the Corrater SCA1 by Kiene and Wable as part of an AwwaRF project (Vasconcelos et al. 1996). The corrosion velocity was measured in a branch of the distribution network in Paris, France. The free chlorine decay rate was also determined on site from knowledge of the water residence time in a pipe section of known diameter. This rate was compared to the theoretical rate calculated from Equation 2.8. The experimental decay rate was about four times smaller than predicted; they attributed the difference to inaccurate measurements of corrosion velocity. The measurements could have been inaccurate because the Corrater SCA1 used electrodes of mild steel while the pipes were cast iron and the electrodes were new while the pipe walls were covered with corrosion products. These field-scale experiments also revealed that the zero-order rate was notably higher at a Reynolds number of 60,000 than at 40,000. While not discussed, it is possible that more turbulent flow increased the availability of Fe^{2+} from tubercles on the pipe.

Electrochemical laboratory-scale experiments were reported by Kiene, Lu, and Levi (1998). Free chlorine was added to a small reactor that contained an iron electrode and a reference electrode. Instead of corrosion rate, the intensity of the corrosion current, i_{corr} (μA) was measured. A linear relationship was obtained between free chlorine decay rate and corrosion current. The same relationship was measured in the presence of 44 mg/L of oxygen (pure oxygen stripping) as in the absence of oxygen indicating that free chlorine is kinetically favored over oxygen for oxidation of Fe^{2+} during corrosion. As had been observed in field-scale experiments by Kiene and Wable (Vasconcelos et al. 1996), Kiene, Lu, and Levi (1998) found that the experimental chlorine decay rate was much greater than the theoretical rate predicted by Equation 2.8; in this work, the discrepancy was a factor of two.

A pilot-scale Propeller Loop Reactor (PLR) was used by Kiene, Lu, and Levi (1998) to explore the relationship between free chlorine decay rate and corrosion rate under more practical experimental conditions. The reactive surface of the PLR is an entire pipe section that forms the outer wall of the bench-scale reactor. An inner cylindrical section of inert material is inserted to form a slot with the outer wall. Water is rapidly mixed by a propeller stirring device within the inner cylinder. While not explicitly discussed by Kiene, Lu, and Levi, the rotational speed of the propeller stirrer controls the water velocity near the pipe wall. The same design has been described by Appenzeller et al. (2001) as the Propella™ reactor. Kiene, Lu, and Levi operated the reactor in batch mode to test the zero-order kinetic model represented by the integration of Equation 2.8 in the form of:

$$C_t = C_o\left(1 - K \times \frac{V_{cor}}{D} \times t\right) \tag{2.10}$$

where

C_o = the initial free chlorine concentration

C_t = free chlorine concentration at any time, t

K = experimentally determined rate coefficient.

The corrosion velocity was measured by a Corrater SCA1 in this study. V_{cor} was increased in a series of experiments by addition of increasing concentrations of NaCl. A linear relationship (R^2 = 0.73) was found between the overall zero-order rate coefficient (KV_{cor}/D) that was obtained from the experimental observation of free chlorine decay and V_{cor}/D from each experiment. The best-fit value of K from this relationship was 1.36×10^{-6} mg/L. This value is about twice the theoretical prediction based on Equation 2.8 but nonetheless is surprisingly close given all the assumptions.

Another laboratory device to measure wall reaction in complete sections of pipe was reported by Hallam et al. (2002). The pipe sections were 100 mm in diameter and 450 mm in length; a magnetic stirrer was used to provide complete mixing. Chlorine wall reaction was interpreted as a first-order rather than zero-order reaction. The wall reaction rate coefficients for cement-lined cast iron were far greater than for plastic pipe. However, the wall reaction rate coefficients varied greatly among the cement-lined cast iron materials (from <0.1 to 0.55 h^{-1}) despite the fact that almost all were new and thus not impacted by long term formation of corrosion products.

Baribeau et al. (2001) used an annular reactor design to study the influence of pipe wall material on chlorine decay. This reactor was originally developed at Montana State University (Characklis and Marshall 1990). An inner PVC cylinder rotates while the outer cylinder remains stationary. Water is passed continually through the reactor. Small coupons of pipe material are attached to the inner cylinder. The annular space between the inner and outer cylinders and the rotational speed are designed to simulate the velocity and shear stress within a distribution system pipe. Residence time is controlled by the water flow rate. While not specifically noted, the units to express reaction rate of chlorine with pipe material reveal that the data were interpreted with a zero-order rate model.

Recent research has shown that monochloramine can react with Fe^{2+} (Vikesland and Valentine 2000). The reaction pathway is very complex and it involves both the bulk solution reaction between monochloramine and Fe^{2+} and the subsequent reaction of Fe^{2+} surface complexes on the ferric oxide precipitate that is deposited from the bulk solution reaction. The net effect of the oxide surface is to accelerate the overall reaction leading to a faster rate of monochloramine decay than can be explained from the bulk solution reaction. This reaction scheme is highly dependent upon the pH. A key intermediate is the amidogen radical ($\cdot NH_2$). It is proposed that the radical may react through pathways that do not involve iron oxidation, particularly at pH of 7 in contrast to pH of 8.5.

Valentine et al. (2000) compared free chlorine and monochloramine decay rates in laboratory experiments with pipe deposit material that was collected from excavated pipe. They noted that the deposits were porous and thus, the interior of deposits could be available for reaction. In experiments with both free chlorine and monochloramine, the surface mediated reaction dominated over the bulk solution reaction. However, the decay rate of free chlorine was substantially higher than that of monochloramine. The decay rate of free chlorine was described as bi-phasic, involving an initial short and fast reaction, complete in less than ten hours, followed by a slower loss that occurred over the remaining exposure time. The initial fast reaction period was attributed to simultaneous reactions in solution and at the deposit surface whereas the

Table 2.2
Comparison of first order coefficients (with 95% confidence intervals) for free chlorine and monochloramine decay in the presence of pipe deposit material (Deposit WA-2) at pH of 7.5

	Concentration of pipe deposit material		
	0.5 g/L	1.0 g/L	2.0 g/L
Free chlorine decay (h^{-1})	0.0048 ± 0.0075	0.0097 ± 0.064	0.034 ± 0.125
Monochloramine (h^{-1})	0.0027 ± 0.0031	0.0061 ± 0.046	0.025 ± 0.27

Data from: Valentine et al. 2000.

subsequent slow reaction period was attributed solely to reactions at the deposit surface. The initial fast reaction period was not observed with monochloramine. A first-order model was used to describe both the free chlorine and monochloramine decay rates. Illustrative results are presented in Table 2.2. The first-order rate coefficients increased in proportion to the weight of deposit material used in each experiment indicative of a surface motivated reaction. The rate coefficients for free chlorine were about twice those for monochloramine but it should be noted that the difference reached about an order of magnitude for other deposit materials.

IRON CORROSION AND DISINFECTION OF BIOFILMS

LeChevallier et al. (1993) examined the extent to which corrosion control could reduce biofilm density. The premise of this research was that a faster corrosion rate produces a faster decay rate of free chlorine or monochloramine within the biofilm on the pipe surface; hence, bacterial inactivation efficiency is reduced. Extensive pipe loop experiments were conducted to measure efficiency of free chlorine and monochloramine in control of biofilm bacteria under different corrosion conditions. Disinfectant decay rates were not measured. Rather, the extent to which biofilm bacteria were inactivated is an indirect measure of the disinfectant decay rate. The corrosion rate was quantified using the Corrater RCS 9000 with carbon steel probes. The findings were consistent with the notion that Fe^{2+} controls the rate of reaction of either free chlorine or monochloramine and that free chlorine is faster reacting than monochloramine. In semi-quantitative terms, if the corrosion rate was above 1 mil per year (mpy) (25.4 μm/year), very little inactivation of biofilm bacteria was achieved by free chlorine despite the introduction of over 4 mg/L into the pipe loop. More biofilm inactivation was achieved by monochloramine than free chlorine at corrosion rates above 1 mpy (25.4 μm/year) implying its slower reaction with Fe^{2+}. However, extremely high corrosion rates also rendered monochloramine ineffective even when introduced at concentrations between 3 and 5 mg/L.

Follow up studies in two distribution systems by Norton and LeChevallier (1997) provided further evidence that corrosion rate was a key determinant of biofilm inactivation by disinfectants. The Hopewell, Va. distribution system was characterized by a high corrosion rate (>4 mpy) (>101 μm/year) and coliform occurrences could not be reduced despite increasing free chlorine concentrations from 0.57 to 2.1 mg/L. Conversion to chloramine reduced coliform occurrences suggesting again that chloramines are less reactive toward Fe^{2+} when the corrosion rate is high. Subsequent corrosion control measures led to even further reductions in coliform occurrences presumably because less Fe^{2+} was available to react with chloramine.

A direct evaluation of free chlorine and chloramine decay through reactions with biofilms was made by scraping biofilms from different pipe surfaces used in pipe loop experiments and measuring the loss of chlorine over a 30-minute period (LeChevallier, Lowry, and Lee 1990). The free chlorine demand exerted by biofilms from cast iron pipe was 40 mg/L compared to 3-4 mg/L for PVC. The greater demand was postulated to originate from accumulation of reduced iron corrosion products within the biofilm. The monochloramine demand (32 mg/L) was lower than free chlorine demand (40 mg/L) for biofilms scraped from cast iron pipe. These investigators reasoned that Fe^{2+} complexes with extracellular polymeric substances (polysaccharides) generated by biofilm organisms and that free chlorine reacts with this material causing liberation of Fe^{2+} that could be subsequently oxidized by free chlorine. They state that monochloramine does not react with this material and thus, more disinfectant is available for inactivation of the biofilm organism.

In a more recent study, Baribeau et al. (2001) deduced the chlorine decay rate caused by biofilms from measurements in full-scale distribution systems. This was accomplished indirectly by comparing the chlorine decay rates in large diameter, concrete pipes (with biofilms but no corrosion by-products present), small-diameter gray iron pipes (with both biofilms and corrosion by-products present) and in bulk water (with no biofilm and no corrosion by-products present). These investigators concluded that the chlorine decay rate caused by the biofilm was much smaller than that caused by the corrosion by-products.

Abernathy and Camper (1998) reported on the effect of corrosion on biofilm inactivation. They used the annular reactor design in which small slides (surface area of about 19 cm^2) of mild steel were placed on the inner cylinder. A continuous flow of water with different feed concentrations of free chlorine was fed to the reactor to produce different free chlorine residuals leaving the reactor. The biofilm density on the slides decreased from the control in which no free chlorine was fed to the dose that produced a residual of 0.1 mg/L in the bulk water. However, biofilm density then increased with free chlorine residual up to 0.75 mg/L and then decreased with higher residuals. One of the reasons cited for the increase in biofilm density was a possible increase in corrosion product reaction rates with new reaction pathways that could consume free chlorine within the biofilm thus negating the effect of higher residuals.

The effect of zinc orthophosphate on corrosion, biofilm density and chloramine residual was also reported by Volk et al. (2000) using the same annular reactor design as used by Abernathy and Camper (1998). The addition of this corrosion inhibitor into the feed for one month gave very little in reduction in biofilm density on cast iron coupons although the corrosion rate was reduced from 4.9 mpy to 2.5 mpy. The chloramine concentration decreased from 3.4 mg/L in the feed to about 2.6 mg/L in the effluent with or without the addition of the corrosion inhibitor. These authors suggest that the experiment was not long enough to achieve benefits. More positive results were reported in a survey of coliform occurrences in distribution systems by LeChevallier, Welch, and Smith (1996). Coliform occurrences were 36% lower in systems using phosphate-containing corrosion inhibitors. Although hypotheses are difficult to prove in full-scale studies, these results suggest that sequestration of corrosion products led to a decrease in disinfectant decay.

Corrosion products may also serve as nutrients to increase biofilm growth in the absence of disinfectant (Appenzeller et al. 2001, LeChevallier et al. 1993). Appenzeller et al. (2001) demonstrated this effect in continuous flow experiments using a section of corroded cast iron pipe in the Propella™ reactor design. Both total concentrations and bacteria counts leaving the reactor were greatly reduced by addition of phosphate for only three days.

MASS TRANSFER RESISTANCE AT PIPE SURFACES

A simple model has been developed to account for mass transfer resistance in transport of disinfectant across the boundary layer adjacent to the pipe wall (Rossman, Clark, and Grayman 1994). The assumption is that the wall reaction rate is first order such that the flux (mass/wall area/time) is:

$$N = k_f(C_b - C_w) = k_w C_w \tag{2.11}$$

where

N = flux of chlorine to pipe wall (mass/area/time)
k_f = mass transfer coefficient (length/time)
C_b = concentration of disinfectant in the bulk
C_w = concentration of disinfectant at the wall
k_w = first-order rate coefficient for wall reaction (length/time)

The rate expression to account for simultaneous bulk and wall reaction within a section of pipe is

$$-\pi r^2 L\frac{dC_b}{dt} = k_b C_b(\pi r^2 L) + N(2\pi r L) \tag{2.12}$$

where

r = pipe radius
L = pipe length
k_b = first-order rate coefficient for bulk chlorine decay (1/time)

Before substituting for flux of chlorine to the wall, N, Equation 2.11 is rewritten to express C_w be in terms of C_b to give

$$C_w = \frac{k_f}{(k_w + k_f)} C_b \tag{2.13}$$

$$N = \frac{k_w k_f}{(k_w + k_f)} C_b \tag{2.14}$$

Equation 2.12 can then be written as

$$-\frac{dC_b}{dt} = \left(k_b + \frac{(2\pi r L)k_w k_f}{(\pi r^2 L)(k_w + k_f)}\right)C_b = \left(k_b + \frac{k_w k_f}{r_h(k_w + k_f)}\right)C_b \tag{2.15}$$

where r_h = hydraulic radius (i.e., $r/2$ for pipes flowing full)

Correlation equations are available in the literature to relate k_f; to water velocity, water viscosity, pipe diameter and diffusivity of the chemical that is transferred across the hydrodynamic boundary layer.

The approach taken by Rossman, Clark, and Grayman (1994) in the development of the EPANET model (Rossman 2000) to determine k_w was to determine k_b for free chlorine from batch laboratory reactor tests and to measure free chlorine concentrations at locations within the distribution system. Given a calibrated hydraulic model for a distribution system, the water quality component can then be added. The reaction rate model for disinfectant decay within EPANET for each pipe section is given by Equation 2.15, requiring that the user enter a value for both k_b and k_w. The bulk reaction coefficient, k_b, is generally specified globally for all pipes. The wall reaction component is then calibrated by adjusting k_w until the observed and predicted concentrations of free chlorine agree within a specified tolerance. In the case history reported by Rossman, Clark, and Grayman, the same k_w was assigned to all pipe sections. These authors acknowledged that finding a single k_w to explain chlorine loss throughout the entire distribution system was simplistic because k_w depends upon pipe materials and their condition. Notwithstanding this simplifying assumption, the calibration results showed that wall reaction accounted for 67% of the chlorine losses in the distribution system compared to only 12% for bulk decay; the remainder of the chlorine loss occurred in tank storage.

An alternate approach to estimation of k_w that does not depend upon the use of a water quality model such as EPANET is to directly calculate wall reaction. This approach has been described as the in-situ method and was used in a recent study by Hallam et al. (2002). The bulk decay rate coefficient, k_b, is measured at a selected point, A, in the distribution system. The travel time between point A and B is estimated from a tracer study or from flow meter records and the decrease in chlorine concentration between points A and B is measured. From this information and the assumption of first-order decay, a wall reaction rate coefficient, k_w can be found from:

$$\frac{C_B}{C_A} = e^{-(k_b + k'_w)t} \tag{2.16}$$

where

C_A = chlorine concentration at Point A
C_B = chlorine concentration at Point B
k'_w = first-order rate constant in units of time^{-1} ($k_w = k'_w r_h$
t = travel time between Points A and B

Hallam et al. measured k'_w values in pipe lengths of cast iron, spun iron, cement-lined ductile iron, two plastic pipe materials (PVC and MDPE). The highest average k'_w values were calculated for the cast- and spun iron (0.67 h^{-1} and 0.33 h^{-1}, respectively), followed by cement-lined ductile iron (0.13 h^{-1}) and then by two plastic materials (0.05 h^{-1} for PVC and 0.09 h^{-1} for MDPE). These results also revealed a fairly strong positive correlation between k_w and water velocity for cast- and spun iron pipes. This would indicate that mass transfer of chlorine to the surface of these materials was the rate-limiting step in chlorine decay rather than the generation rate of corrosion products.

An alternative to Equation 2.15 is needed if the wall reaction is zero- instead of first-order, which would be true of the iron corrosion reaction with chlorine given by Equation 2.8. In this instance, the reaction rate is limited by the rate of generation of iron corrosion products at the surface rather than by the concentration of chlorine. However, if the wall reaction rate is fast relative to the mass transfer rate of chlorine to the surface such that chlorine is immediately consumed at the wall, the reaction would then be limited by mass transfer of chlorine from the bulk water. Vasconcelos et al. (1997) proposed the following to account for these two possibilities:

$$-\frac{dC_b}{dt} = k_b C_b + \min\left(\frac{k_w}{r_h}, \frac{k_f}{r_h} C_b\right) \tag{2.17}$$

This equation states the smaller of the two rates (intrinsic wall reaction vs. mass transfer rate of chlorine to the wall) controls the observed wall reaction. If the wall reaction is slow relative to the mass transfer rate, then the rate is independent of the disinfectant concentration reaching the wall as a consequence of the zero-order nature of the wall reaction. Vasconcelos et al. suggested that zero-order wall reactions could be associated with free chlorine decay at the surface of corroded cast iron pipe as has been noted in other laboratory studies (Frateur et al. 1999 and Kiene, Lu, and Levi 1998) and in distribution system studies (Vasconcelos et al. 1996).

A survey of the literature resulted in a listing of k_b values for free chlorine from various studies that is presented in Table 2.3. The k_b values range from 0.002 to 0.74 h^{-1} and illustrate that the reaction environment can vary widely among distributions systems.

A survey of the literature was also conducted for the wall reaction coefficient. This revealed two different interpretations of kinetics (first- vs. zero order). The variation in k_w is large from site to site as may be expected because age and type of pipe material, as well as bulk water quality, directly impact the wall reaction. A listing of first-order decay coefficients is provided in Table 2.4 from various studies.

EPANET was used to calibrate k_w in four distribution systems (Vasconcelos et al. 1997). Both first- and zero-order rate models were fit to the same chlorine residual data for several distribution systems using EPANET. Three calibration procedures were used to find the chlorine decay coefficient (either first- or zero-order rate coefficient) that gave the best fit between predicted and observed field measurements: (1) all pipes assigned the same best-fit k_w (global method), (2) pipes in specific sections assigned the same best-fit k_w (zonal method), and (3) calculating the best-fit k_w for each pipe. The assumption each method is that k_w is inversely proportional to the pipe roughness coefficient (Hazen-Williams C-factor):

$$k_w = \frac{\alpha}{CF} \tag{2.18}$$

where

α = roughness constant
CF = C-factor in Hazen-Williams equation

Equation 2.18 shows that the lower the C factor, the higher is k_w consistent with rougher pipe surfaces causing greater wall demand. An automated calculation procedure is provided within EPANET by which the user enters a trial value of α and predictions of chlorine residual are compared with field measurements. The α is then adjusted to achieve the best fit of the data to the model predictions. All three approaches gave reasonably good fits to the chlorine residual data (absolute error in the range of 0.05 to 0 0.15 mg/L and relative error between 17 and 31 percent).

Table 2.3
Example literature values of first-order rate coefficients for free-chlorine decay in bulk water

System	k_b (d^{-1})	Reference
Bellingham, Wash.	1.46	Vasconcelos et al. (1997)
Fairfield, Calif.	2.00	Vasconcelos et al. (1997)
North Marin, Calif.	2.29-30.83	Vasconcelos et al. (1997)
Harrisburg, Penn.	0.42	Vasconcelos et al. (1997)
North Penn, Penn.	0.13- 1.33	Vasconcelos et al. (1997)
Severn Trent region, UK	0.42-13.75	Powell et al. (2000)
South Central Connecticut Regional Water Authority, Conn.	0.96	Rossman, Clark, and Grayman (1994)
Pilot-scale	0.13	Li et al. (2000)
Bottle test	2.50-22.50	Kiene, Lu, and Levi (1998)
Seine river treatment plant, France	1.25	Kiene, Lu, and Levi (1998)
Ohio River water, Cincinnati, Ohio	0.25-0.88	Rossman et al. (2001)
Eagle Reservoir, Ind.	0.50-1.38	Rossman et al. (2001)
Bottle test	0.521	Maier, Powell, and Woodward (2000)
Bottle test	0.42-7.92	Hua et al. (1999)
Manchester, Conn.	0.08-0.13	LeCourt et al. (1999)

However, the values of k_w varied widely from system to system as may be expected because age and type of pipe material directly impact the wall demand. The global first- and zero-order rate coefficients, k_w and K_w, respectively, and the α values are listed in Table 2.5. For use of the C-factor approach, α varied widely from system to system given that it was the calibration parameter from which k_w were calculated. The variation in wall demand could be due to differences in pipe material, pipe as described earlier.

PIPE LOOPS TO EVALUATE DISINFECTANT DEMAND

Pipe loops provide a larger-scale alternative to annular reactors and the Propella™ reactor as described earlier. The advantage is that the test condition can simulate the effects of pipe material and water velocity in a more realistic manner. While the ideal simulation of the reaction environment in the pipe would be to waste test water continuously as occurs in actual operation, practical considerations often require that a batch of test water be recirculated, especially if test waters must be transported to the laboratory.

Table 2.4
Example literature values of first-order coefficients for free-chlorine wall reaction

System	k_w or k_w/r_h	Pipe material	Pipe diameter (in.)	Reference
South Central Connecticut Regional Water Authority, Conn.	0.15-0.45*		8-12	Rossman, Clark, and Grayman (1994)
Bellingham, Wash.	0.76*	Unlined CI, > 40 yr	6-8	Vasconcelos et al. (1997)
Fairfield, Calif.	0.0*	Asbestos-cement pipe, <15 yr	8-12	Vasconcelos et al. (1997)
North Marin, Calif.	1.52*	Unlined CI >40 yr	12-30	Vasconcelos et al. (1997)
Harrisburg, Penn.	0.27*	Unlined galvanized iron or steel, 30-50 yr	2-6	Vasconcelos et al. (1997)
North Penn, Penn.	0.03*	Cement-lined DI and unlined CI	8-16	Vasconcelos et al. (1997)
Manchester, Conn.	0.046-0.091*		4-12	LeCourt et al. (1999)
Pilot-scale	0.14-2.88†	Synthetic pipes		Kiene, Lu, and Levi (1998)
Pilot-scale	3.84†	Gray CI		Kiene, Lu, and Levi (1998)
Pilot-scale	2.03-4.69†	Unlined DI	6	Rossman et al. (2001)
Pilot-scale	1.53-3.57†	Unlined DI	6	Rossman et al. (2001)
Pilot-scale	2.10-3.15†	Unlined DI	6	Rossman et al. (2001)
Cincinnati, Ohio	3.8-10†	Unlined DI, 70-110 yr old		Vandermeyden and Hartman (2001)

* Hydraulic radius not included in the rate coefficient (m/d)
† Hydraulic radius included in the rate coefficient (1/d)

Table 2.5
Comparison of first-order and zero-order wall reaction from EPANET calibration in five distribution systems

	First-order model		Zero-order model		
Site	k_w (m/d)	α	K_w (mg/m^2/d)	α	Percent of pipes with wall reaction
Bellingham, Wash.	0.76	61	215	17,216	95
Harrisburg, Penn.	0.27	24.4	91.5	6,994	100
North Marin, Calif.	1.52	198	215	27,976	13
North Penn, Penn.	0.03	3.05	53.8	5,380	100

Data from: Vasconcelos et al. 1997.

Table 2.6
Comparison of first-order rate coefficients for bulk and wall demand for free chlorine in pipe loop experiments

Test water	Average k_b (d^{-1})	Average k_w/r_h (d^{-1})
CWW-C*	0.266	3.44
INDY-D†	0.492	2.72
CWW-E‡	0.251	2.45

Data from: Rossman et al. 2001.
*Batch C of Settled-filtered (unchlorinated) water from the Cincinnati, Ohio Water Treatment Plant; free chlorine added to batch at beginning of pipe loop test.
†Settled-filtered (unchlorinated) water from Eagle River reservoir, Indianapolis, Ind.; free chlorine added to batch at beginning of pipe loop test.
‡Batch E of settled-filtered (unchlorinated) water from the Cincinnati, Ohio Water Treatment Plant; free chlorine added to batch at beginning of pipe loop test.

Rossman et al. (2001) reported on a set of pipe loop tests that give insight into the importance of wall reaction. The pipe loop consisted of a 27 m (88 ft) length of 150 mm (6 in) diameter unlined ductile iron pipe through which water passed at high enough velocity to eliminate mass transfer resistance at the pipe wall. For this condition, Equation 2.15 reduces to:

$$-\frac{dC_b}{dt} = \left(k_b + \frac{k_w}{r_h}\right)C_b = KC_b \tag{2.19}$$

where K = overall first-order rate coefficient

Equation 2.18 can be integrated for the batch condition provided in the pipe loop. Separate bulk water tests of chlorine decay were used to measure k_b such that k_w could be deduced from the overall rate coefficient in the pipe loop experiment. The resulting average values of k_b and k_w/r_h (both in the same units of d^{-1}) for three water sources are given in Table 2.6. For each test water, the wall reaction rate coefficient was about an order of magnitude greater than the bulk decay rate coefficient, again emphasizing the importance of wall reaction when corrosion is important. Rossman et al. noted that k_w/r_h generally decreased in six batch tests conducted over a three-month period. They attribute the decrease in decay rate to a possible depletion of chlorine-demanding substances attached to the pipe wall, although acclimation of the pipe material to higher chlorine levels than previous exposure is also cited.

A modification of the pipe loop test was described by Valentine et al. (2000). Instead of using a section of pipe material, a piece of pipe deposit material that had been removed intact from a distribution system pipe was inserted into a section of glass tubing. A batch of ultrapure (disinfectant demand free) water dosed with a disinfectant was circulated through the tubing and the disinfectant (free chlorine or monochloramine) decay rate was measured. The same rate of free chlorine decay was measured for several experiments in which the deposit material composition was notably different. The explanation proposed was that the observed rate of free chlorine decay was influenced by mass transfer at the deposit surface and that both exterior and interior surfaces could be involved. Evidence for external mass transfer control was provided by

Table 2.7
First order decay rate coefficients in modified pipe loop test using single piece of pipe deposit material (2.14 g of CR-2 sample); pipe loop volume = 550 mL, nominal pH = 6.5, dissolved oxygen = 8.0 mg/L

Spike*	First-order decay coefficients (h^{-1})
1	0.061
2	0.111
3	0.112
4	0.059
5	0.042
6	0.054

Data from: Valentine et al. 2000.
*Initial free chlorine concentration of 10 mg/L as Cl_2 and repeated spikes with same deposit material.

observing that the free chlorine decay rate decreased in proportion to the decrease in flow rate through the test loop. A rate model that includes mass transfer in series with a first-order surface reaction was proposed (Equation 2.15) where r_h is inverse of the surface area of deposit-to-volume of deposit. Illustrative experimental values of K are listed in Table 2.7. Two of these were in the range of first-order coefficients for wall reaction found by Rossman et al. (2001) in pipe loop experiments and shown in Table 2.7 (0.10-0.14 h^{-1}) while the rest were about 50% lower.

The chlorine demand of pipe deposits was also investigated by Gauthier et al. (1999). Loose deposits were obtained from the DS in Nancy, France. A carbon-to-nitrogen ratio of about seven suggested a strong influence of in-situ biological activity. The chlorine demand (mg chlorine per g of dry deposit) increased from about 10 to 100 as the organic carbon content increased from 2 to 8 percent. Very high bacterial counts, ranging from 1×10^8 to 5×10^8 colony forming units per g of deposit, were present owing to the loss of chlorine residual by reaction with the deposit.

The effect of dissolved oxygen on free chlorine decay rate was also investigated by Valentine et al. (2000) using the same experimental set up. Reducing the dissolved oxygen to less than 2 mg/L produced a significantly slower rate of free chlorine decay (0.012 h^{-1}) than was observed in the presence of 8 mg/L of dissolved oxygen (0.053 h^{-1}). This result is contrary to the expectation that lowering the dissolved oxygen concentration would decrease the shell-like protective layer of oxidized iron and release more Fe^{2+} (Sarin et al. 2001) thus increasing the reaction rate with free chlorine. Valentine et al. hypothesized that the decrease in decay rate at low oxygen level is caused by the absence of the superoxide (O_2^-) radical. At high oxygen levels, the oxidation of Fe^{2+} to Fe^{3+} by free chlorine could lead to production of O_2^- from dissolved oxygen and this could catalyze the reduction of additional free chlorine thereby accelerating the chlorine decay rate. However, O_2^- would not be abundant at low oxygen levels and thus, this catalytic pathway for chlorine reduction would no longer be important. No evidence was provided to support this hypothesis.

Valentine et al. (2000) also found that the observed first order decay rate for monochloramine was an order of magnitude lower than for free chlorine in the same modified pipe loop experiments. They noted that the rate coefficients for both free chlorine and monochloramine decay decreased upon repeated spikings of either disinfectant over 1500 hours. While the total disinfectant demands were rather large (7.3 mg free chlorine/g and 2.8 mg chloramine as Cl_2/g), this suggested that the pipe deposit material has a finite capacity to react with either disinfectant.

Pipe loop experiments were also conducted by Abernathy and Camper (1998) to investigate the impact of free chlorine or monochloramine on biofilm density as well as the impact of zinc orthophosphate and polyphosphate corrosion inhibitors. Substrate and nutrients were added to drinking water from Bozeman, Mont. that had been further processed to remove biodegradable substrate and residual free chlorine. Continuous subsequent feeding of 1.1 to 1.5 mg/L of free chlorine or 0.7 mg/L of monochloramine into the mild-steel pipe loop did not reduce the established biofilm density significantly, which is consistent with the idea that both disinfectants are consumed by reactions with corrosion products; the free chlorine and monochloramine residuals were very low (0.05 and 0.02 mg/L, respectively). Although these authors concluded that corrosion control would increase disinfection efficiency and lower biofilm densities, their results suggest only modest benefits. The addition of either zinc orthophosphate or polyphosphate to the chlorinated feed did not reduce biofilm density although the free chlorine residual was somewhat higher (0.1 mg/L compared to 0.05 mg/L); thus, free chlorine decay still occurred due to reactions at the pipe surface. When these corrosion inhibitors were introduced with monochloramine, there was only a very modest reduction in biofilm density and the monochloramine residual was somewhat higher (0.1-0.2 mg/L compared to 0.02 mg/L without the inhibitor).

CHAPTER 3
METHODS AND MATERIALS

PIPE-SECTION REACTOR EXPERIMENTS

The overall goal of this part of the research was to design, construct and test a new type of laboratory reactor for study of wall reaction kinetics. The low-cost, easy-to-operate new reactor included an entire pipe section and allows for study of the effects of velocity and residence time. Soon after the pipe-section reactor was built, a similar device (the propeller loop reactor) was found in the literature (Kiene, Lu, and Levi 1998). This reactor was developed in France at CIRSEE in 1978; a detailed analysis of hydrodyanamics was provided Blenke (1978). In the experiments described by Kiene, Lu, and Levi (1998), the reactor wall was made of different types of plastic pipe materials to isolate the effect of total organic carbon and temperature on bulk demand. Additional studies were made of the effects of biofilm development and corrosion rate on bulk and wall demand. However, the propeller loop reactor did not provide a way to control the water velocity in the pipe section. The specific objectives of this current research, which were not included in the Kiene, Lu, and Levi (1998) study, were to quantify the effects of water velocity and water quality (corrosion rate, dissolved oxygen, and pH) on chlorine wall demand in DI- and CI pipes.

Reactor Design

The design of the pipe-section reactor is shown in Figure 3.1. The outer cylinder is the intact section of pipe material to be studied and the inner cylinder is made of Plexiglas. A large, propeller-type stirrer is used to induce movement of the water from the inner cylinder to the annular space between the inner and outer cylinder. The rotational speed of the stirrer motor was adjusted by a rheostat and this produced a range of water velocity in the annular space (see later section for discussion of Velocity Measurements). Two sets of four plexiglas baffles, spaced at 90 degrees around the cylinder wall, were attached to the inner cylinder to prevent vortexing. The simulated water velocity in this annular space is equivalent to the velocity of interest near pipe walls.

Both continuous flow and batch modes of operation can be used to investigate the kinetics of wall demand. In the continuous flow mode, the reactor behaves as a completely-mixed flow reactor (CMFR) due to the intense stirring rate. In the batch mode, the reactor behaves as a completely-mixed batch reactor (CMBR). In either mode, experiments can be designed easily to test different wall reaction rate models. As a CMBR, it should be noted that chlorine decay as a function of time would in theory be identical to that observed in the same diameter pipe in a distribution system as a function of water age. In either mode, the pipe-section reactor provides a more realistic simulation of conditions within a pipe than in the annular reactor design where only a small coupon of pipe material is used.

During kinetic experiments using chlorine, the outer cylinder was either new DI or old CI pipe taken from a distribution system. The dimensions of each pipe section and the volume of water within the reactor are given in Table 3.1. A continuous flow of water initially passed through the reactor at high rate to measure the corrosion rate and to make sure that the chlorine residual was stable. The water flow was then stopped and the chlorine decay experiment was conducted with the reactor operating as a CMBR.

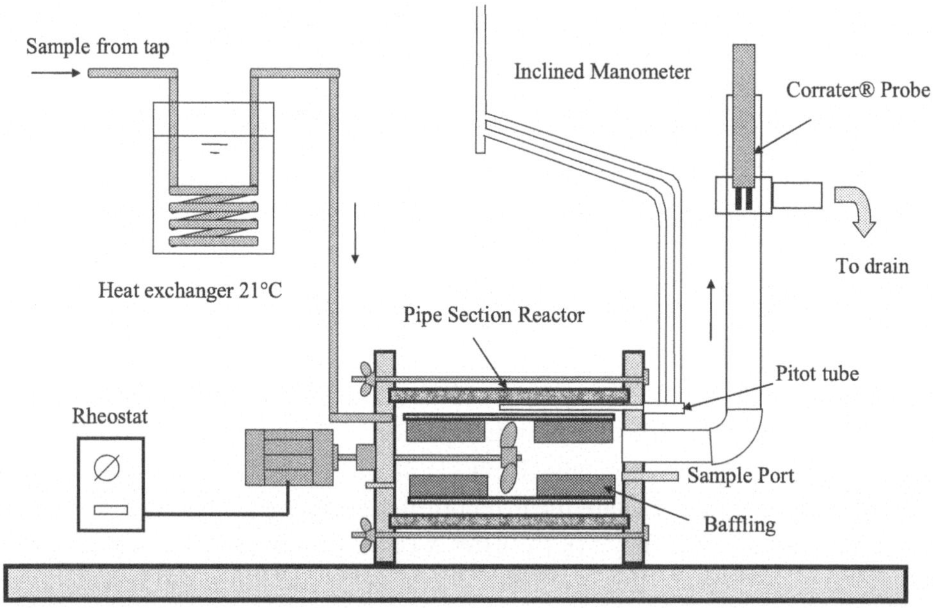

Figure 3.1 Schematic of pipe-section reactor (cut-away section showing inner Plexiglas cylinder and two sets of four baffles; four others not shown are spaced at 90 degrees)

Table 3.1
Materials and dimensions of pipe-section reactor

	Outer diameter, D_O (cm)	Inner diameter, D_I (cm)	Length (cm)	Approximate volume within cylinder (L)
DI pipe outer cylinder	17.2	15.7	38.3	6.8
CI pipe outer cyliner	17.1	15.2	38.8	6.7
Plexiglas outer cylinder	15.0	13.8	38.9	5.4
Plexiglas inner cylinder	11.3	10.5	33.8	2.9

Test Water

The reactor was fed with the tap water distributed by the Orange Water and Sewer Authority (OWASA) that provides about 10 mgd (3.785×10^4 m^3/d) to the Carrboro-Chapel Hill community. The raw water derives primarily from two surface water impoundments: University Lake and Cane Creek Reservoir. The average annual alkalinity, pH and TOC of finished water in 2001 were 3.1 mg/L as $CaCO_3$, 7.4 and 3.2 mg/L, respectively. OWASA doses a phosphate-based corrosion inhibitor (Calciquest US2 product which is 18 percent Ortho and 18 percent poly phosphates) at an average annual concentration of 1.3 mg/L. Except in experiments in which dissolved oxygen (DO) was removed, the DO was between 7 and 8.5 mg/L in the test water within the reactor at the beginning of each experiment.

Pipe Material

Six-inch (152 mm) diameter sections of CI and cement-lined DI pipe were provided by OWASA. The DI pipe was from the OWASA supplies and had never been used. The CI pipe was installed in the 1930's and upon its removal in 2000 was heavily corroded with many tubercles on the inner surface. The two pipe samples were cut into 20-inch (508 mm) long sections with relatively little damage to either cement lining of the DI pipe or the tubercles of the CI pipe. The exposed metal surfaces at the both ends of each section were coated with epoxy to prevent exertion of chlorine demand. The pipe sections were re-wet by soaking in tap water for three weeks before assembling the reactor. After installation, tap water (with chlorine residual about 0.7 mg/L) was then run continuously for another one week to flush out any easily dislodged tubercles and to eliminate the chlorine demand of plasticizers in the Plexiglas.

Water Quality Measurements

When the reactor was operating in batch mode, water samples were removed from the reactor by injecting 25 mL of tap water at one end of the reactor with a syringe while collecting the displaced water in a glass sampling cell at the other end. The chlorine residual was measured by the DPD colorimetric procedure using Hach chemical pillows and a Hach® DR2000 spectrophotometer (Loveland, CO). Water temperature and pH were measured with a Fisher Scientific, Accumet® portable AP 61 with Accumet® Ag/AgCl probe. DO was measured using an YSI Model 54A Oxygen meter. After each test run, tap water was passed slowly through the reactor to obtain a measurement of the corrosion rate at the reactor outlet. The corrosion rate was measured using the AquaMate Portable Corrater® with the mild-steel 6112 Corrater® Probes (Rohrback Cosasco Systems, Inc., Santa Fe Springs, Calif.). Corrosion rate is expressed in units of mpy, i.e., the annual loss rate of pipe wall thickness (mpy) (0.025 mm/y). Preliminary experiments showed that a continuous flow of water must pass by the Corrater® Probe for the measurement to be accurate.

Experiments

The decay rate of chlorine in bulk water was measured by placing 2 L of OWASA tap water into a 2-L clean glass bottle in a dark place at room temperature (21°C). Each measurement of chlorine required 25 mL; the sample volume was thus reduced by about 200 to 300 mL over 72 hours. Whether chlorine volatilized into the added headspace to contribute to measurement of chlorine loss is unknown. However, the bottles were not shaken and so, the rate of mass transfer should have been very low.

All experiments with the pipe-section reactor were done in the CMBR mode. The experiments with the CI pipe-section reactor measured the effects of initial chlorine concentration, water velocity, pH, Cl^- and DO on the rate of chlorine decay. Techniques for isolating the wall reaction rate will be discussed under Results. The chlorine concentration in OWASA tap water was increased by addition of about 50-mL of diluted bleach solution (Chlorox®) into the pipe-section reactor to produce initial chlorine concentrations of 1.42 and 2.1 mg/L.

The effects of water velocity, pH, chloride (Cl^-), and DO on chlorine wall demand were also measured. The stirring speed of the propeller mixer in the pipe-section reactor (Figure 3.1) was varied to yield velocities of 0.16 m/s, 0.35 m/s and 0.62 m/s (see Velocity Measurements). A dilute HCl solution was added to produce pH values of 6.6 and 3.6. In another set of experiments, NaCl was added to produce concentrations of 20, 40, 80 and 160 mg/L above the background level to raise the corrosion rate. DO was decreased in two other types of experiments. The first method was to circulate a batch of OWASA tap water in the pipe-section reactor for three days until the DO decreased from 8.2 mg/L to 1.1 mg/L and then to add chlorine for measurement of the decay rate. The second method was to strip the DO from the water with nitrogen gas to reach 1.7 mg/L within 2 hours and once again to add chlorine for measurement of the decay rate. In both types of experiments, the chlorine decay rate was measured at the minimum water velocity that could be accurately determined (0.16 m/s) to simulate dead-end conditions in a distribution system.

Velocity Measurements

Velocity was measured indirectly from the differential manometer height in the Pitot tube that was installed in the annular space between the outer and inner cylinders (see Figure 3.1). The end of the Pitot tube was placed about half way along the length of the pipe section to avoid disturbances of the flow stream created at either end of the inner cylinder. A calibration curve between velocity and differential manometer height was generated from flow rate tests in a straight section of 1-inch diameter PVC pipe. The Pitot tube was installed on the PVC pipe and the differential manometer height was measured at various flow rates. From the flow rate and pipe diameter, the average water velocity in the cross section could be calculated. The resulting calibration curve is shown in Figure 3.2. This relationship agrees with theory for relating head loss to velocity in a flowing pipe (Olson and Wright 1990):

$$v(\mathrm{m/s}) = \sqrt{2 \cdot g \cdot h} = \sqrt{2 \times 9.8/1000 \times \mathrm{h(mm)}} = 0.14 \cdot \sqrt{h} \qquad (3.1)$$

where

v = velocity
g = acceleration of gravity
h = head loss

Water velocities in the range of 0.5 to 2.0 m/s could be measured using the Pitot tube.

The water velocity in the pipe-section reactor fitted with the Plexiglas outer cylinder was calibrated against the rheostat motor setting that controlled the speed of propeller stirrer. The results are given in Table 3.2. Water velocity was also estimated by observing the movement of a neutrally buoyant, polyurethane foam ball with a 5 mm diameter after its introduction into the pipe-section reactor. The ball was assumed to follow the same flow pattern as the water. The velocity of the water was estimated by measuring the travel distance and the travel time of the bead in the reactor. These results are compared with the Pitot tube method in Table 3.2 and they agree fairly well. The Pitot tube method was selected for use in all subsequent tests.

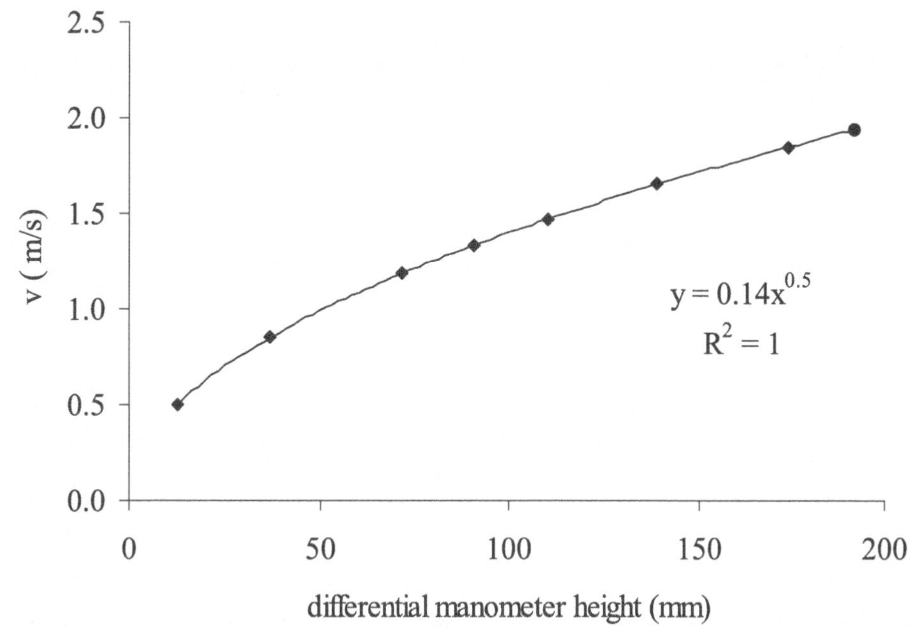

Figure 3.2 Pitot tube calibration curve using a straight section of 1-inch diameter PVC pipe

The following relationship was developed between mixer rheostat power setting and water velocity for the Plexiglas section with a 3-inch propeller stirrer:

$$v_{Pitot}(\text{m/s}) = 0.0021 \times (PS_{3''\text{-}D})^{1.3} \qquad (3.2)$$

where

v_{Pitot} = velocity measured by the Pitot tube

$PS_{3''\text{-}D}$ = power setting of the rheostat to control the stirrer motor for the 3-inch diameter propeller.

As shown in Table 3.1, when either the DI or CI pipe section was used as the outer cylinder of the reactor, a larger water volume was held within the reactor than that when Plexiglas was used. This larger volume required replacement of the 3-inch propeller with a 4-inch propeller to keep the range of water velocity in the typical range of practice. With the 4-inch diameter propeller, the relationship of power setting to velocity was:

$$v_{Pitot}(\text{m/s}) = 0.00039 \times (PS_{4''\text{-}D})^{2.0} \qquad (3.3)$$

Table 3.2
Comparison of measurements of water velocity in the pipe-section reactor with 3-inch diameter propeller by Pitot tube (v_{Pitot}) and foam ball tracer (v_{ball})

	Pitot tube method		Polyurethane foam ball* (3 trials)					
Power setting	Pitot tube differential height (mm)	v_{Pitot} (m/s)	Stopwatch reading 1 (s)	Stopwatch reading 2 (s)	Stopwatch reading 3 (s)	Average travel time (s)	v_{ball} (m/s)	Ratio v_{ball}/v_{Pitot}
30	1.5	0.17	2.09	2.16	2.13	2.13	0.16	90%
40	4	0.28	1.72	1.69	1.72	1.71	0.19	69%
50	9	0.42	0.97	1.09	0.97	1.01	0.33	78%
60	11	0.46	0.91	0.94	0.87	0.91	0.36	78%
70	13	0.51	0.65	0.72	0.69	0.69	0.48	95%

* length of inner pipe = 0.33 m

FIELD MEASUREMENTS

North Chatham County, N.C.

The North Chatham County water distribution system delivers approximately 1 mgd (3.78×10^3 m^3/d) to users in the county, which is located to the south of Chapel Hill. The water treatment plant (WTP) is located at the south end of the system, as shown on Figure 3.3. Raw water is pumped from Jordan Lake, a U.S. Army Corps of Engineers flood control impoundment; this source has a very low alkalinity (<20 mg/L as $CaCO_3$) but high TOC (6-10 mg/L). The treatment process consists of alum coagulation, sedimentation, dual media filtration and disinfection with free chlorine. Because of the rural nature of this system, there are many long pipelines with few user connections. These pipelines were ideal for corrosion measurement because they provide a way to quantify chlorine decay in a pipe without confounding effects from mixing.

Field samples were collected from the distribution system on October 26, 2000 and November 22, 2000. Although separated by only one month, the water temperatures were noticeably different (approximately 20°C on October 26 vs. 14°C on November 22). Sampling locations are shown on Figure 3.3. Samples were collected of the finished water leaving the WTP (Location 1) and at four locations (hydrants) in the northern end of the distribution system. Location 2 is on Farrington Road, 5.3 mi (8.5 km) from the WTP. The pipeline connecting Location 2 to the treatment plant is a 16-in. (406 mm) ductile iron transmission line that is approximately 10 years old. There are few users along this pipeline and no connections with other pipes in the system. Locations 3 and 4 are located 12.8 mi (20.5 km) and 13.8 mi (22 km) from the WTP on US Highway 15-501.

Water can reach locations 3 and 4 by several pathways, including passage through one or more elevated storage tanks; thus, water quality is affected by water storage as well as passage through different long pipe sections. Location 4 is at the northern-most end of the distribution system on US Highway 15-501 at the county line. This is a dead end location and there are very

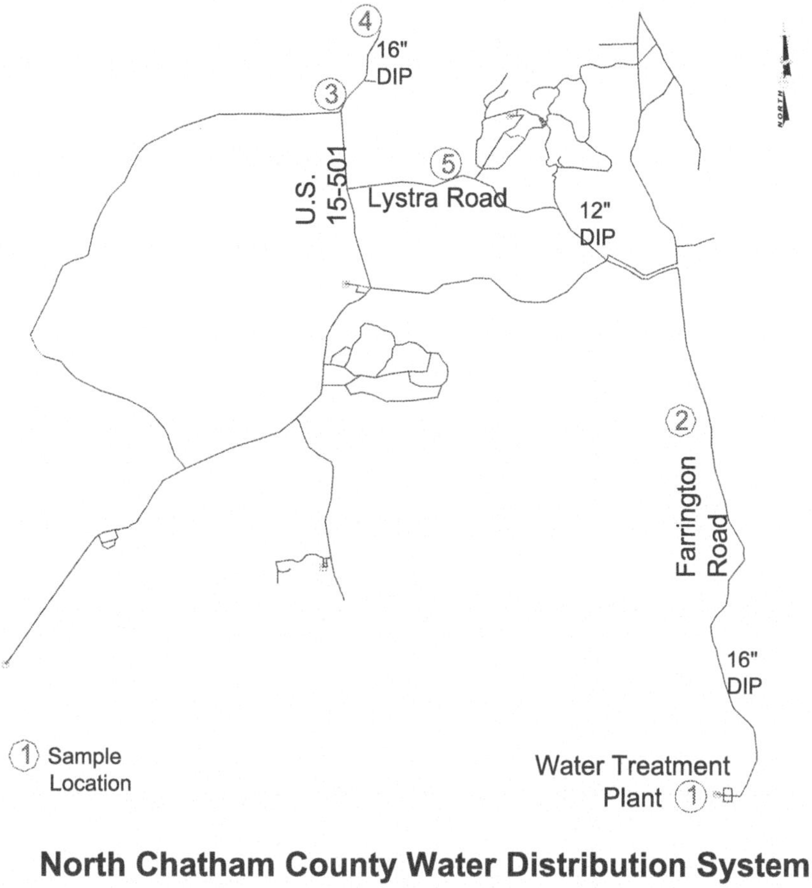

Figure 3.3 Location of sampling sites North Chatham County system

few users between locations 3 and 4. Location 5 is mid-way through the system between locations 2 and 3.

On-site measurements were made of free chlorine (Hach DPD colorimetric procedure and a Hach® DR2000 spectrophotometer), total iron (Hach phenanthroline colorimetric procedure and same spectrophotometer as for free chlorine), temperature, and pH (Fisher Scientific Accumet® portable AP 61 with Accumet® Ag/AgCl probe) and corrosion rate (AquaMate Portable Corrater® and the 6112 Corrater® Probe). Each hydrant was flushed for five minutes after which free chlorine, total iron, temperature and pH were measured. A 3 ft (91 cm) section of 1 in (25.4 mm) diameter PVC pipe was then connected to the hydrant. Corrosion rate measurements were made by inserting the Corrater® probe into the PVC pipe. Measurements were made at five minute intervals and found to be stable after 30 minutes. The chlorine wall demand was calculated for the section of pipe between Locations 1 and 2. The flow in the pipe during both sampling periods was 0.95 mgd (3.6×10^3 m^3/d). Plug-flow conditions were assumed in order to estimate the residence time between Locations 1 and 2. Equation 2.16 was used to calculate k_w from measurements of free chlorine concentration at Locations 1 and 2; k_b was found from measurements of chlorine decay in a batch sample withdrawn from Location 1.

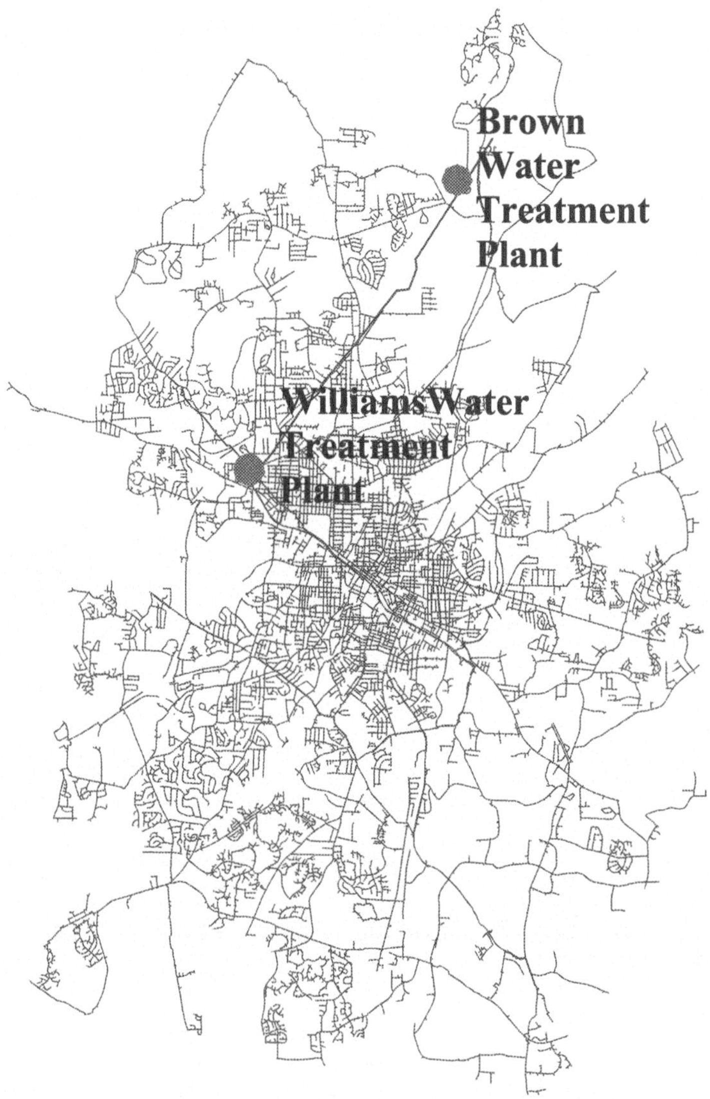

Figure 3.4 Layout of the Durham distribution system

Durham, N.C.

The daily average water demand in Durham is approximately 32 mgd (121×10^3 m^3/d). Raw water is taken from Lake Michie (on the Flat River) and Little River Reservoir. The raw water is characterized by TOC concentrations of 4 to 5 mg/L, color of 40 color units, pH of 6.5 to 7, and alkalinity and hardness of less than 20 mg/L as $CaCO_3$ (Smith 1994). These two raw waters are blended together and fed to the two treatment plants (Williams and Brown) that use conventional water treatment (alum coagulation, sedimentation, chlorine addition before and after dual-media filtration) and a zinc phosphate-based agent for corrosion control.

The Durham DS is shown in Figure 3.4. Pitometer Associates used EPANET to develop a network model (Pitometer Associates 2001). The model contains 11,712 pipe segments with a total length of 1,034 miles, representing all pipes 6 inches in diameter and greater, as well as several hundred pipe segments 4 inches in diameter or less. An extended-period simulation of average day demand in the year 2000 was made in a separate contract from Pitometer Associates to UNC. Chlorine residual data (50 grab samples) were collected during a 5-day period in July 2001 to develop and calibrate the model. Routine sampling done by the Durham staff provided an additional 32 samples. Free chlorine concentration was measured by the DPD method and a portable Hach spectrophotometer model DR/2000a after 5 minutes of flushing.

The sampling plan was developed in consultation with City staff. Locations were selected with low chlorine residuals and these were sampled at least once during the 5-day sampling period. Other samples were taken to provide full coverage of service area as well as to include a variety of pipe materials and sizes. Samples were collected from hydrants (28 out of 50) and at customer taps (22 out of 50). They included three dead ends, ten pipes with diameter of 6 inches or less and seven cast iron pipes. Some samples were also collected from the City's regular sampling sites. Finished water samples were collected from both WTPs to measure chlorine decay in bulk water for calculation of $k_{b.}$

The 72-hour extended period simulation covered the Tuesday through Thursday. Free chlorine data that were collected on Monday and Friday were used to help initialize concentrations throughout the system. The average-day hydraulic model provided by Pitometer Associates was calibrated for the 72-hour period to match four tank levels and flows recorded at each plant. Hazen-Williams C-factors were assigned by Pitometer Associates and were not changed for this calibration. Demand factors and tank controls were adjusted by trial and error to obtain a good replication of tank levels.

The k_w value was calibrated using a globally applied relationship to the pipe roughness coefficient (Hazen-Williams C-factor) available within EPANET. An automated calculation procedure is provided by which the user enters only the roughness constant to relate the C-factor to wall reaction and k_w is calculated for each pipe (Rossman 2000) as described by Equation 2.18. The calibration was further improved by adjusting k_w values in specific pipe segments in the vicinity of sample locations. In some locations, k_w for pipes upstream of sampling location was also adjusted. This technique required hundreds of iterations.

CHAPTER 4
RESULTS

PIPE-SECTION REACTOR

Bulk Decay Rate of Chlorine

The results in Figure 4.1 indicate that a first-order reaction model fit the bulk chlorine decay data for the OWASA water used in the experiments very well over a 72-hr period of observation. The first-order rate model was also tested in a series of 16-hr experiments. The experimental conditions and resulting first-order rate coefficient, k_b, for each experiment are listed in Table 4.1. The mean and standard deviation of k_b was 0.033 ± 0.005 h^{-1}. Additional bulk decay tests of shorter duration were also conducted later in the study and these gave similar results.

Chlorine Decay Rate in CI Pipe-Section Reactor

A series of experiments using CI pipe in the reactor were conducted to determine the wall reaction rate. The chlorine decay pattern is shown in Figure 4.2 for CI pipe section tests performed with three different initial concentrations of chlorine. The linear decline in chlorine residual indicates a zero-order rate process:

$$\frac{dC}{dt} = -\frac{K_w}{r_h} \qquad (4.1)$$

where

C = free-chlorine concentration
K_w = zero-order rate coefficient
r_h = hydraulic radius of pipe section

The integrated form of Equation 4.1 describes the data in Figure 4.2:

$$C = C_o - \left(\frac{K_w}{r_h} \cdot t\right) \qquad (4.2)$$

where

C_o = initial free-chlorine concentration

The slope of each line in Figure 4.2 yields about the same value of K_w (36.5, 34.2, 31.9 mg/m^2/h) which provides confirmation of zero-order kinetics. In this case, the zero-order rate model is admittedly an approximation because the loss of chlorine by first-order decay rate in the bulk water (Figure 4.1) has not been subtracted from the total decay rate. However, the rate disappearance of chlorine is dominated in this case by wall reaction. The magnitude of the wall

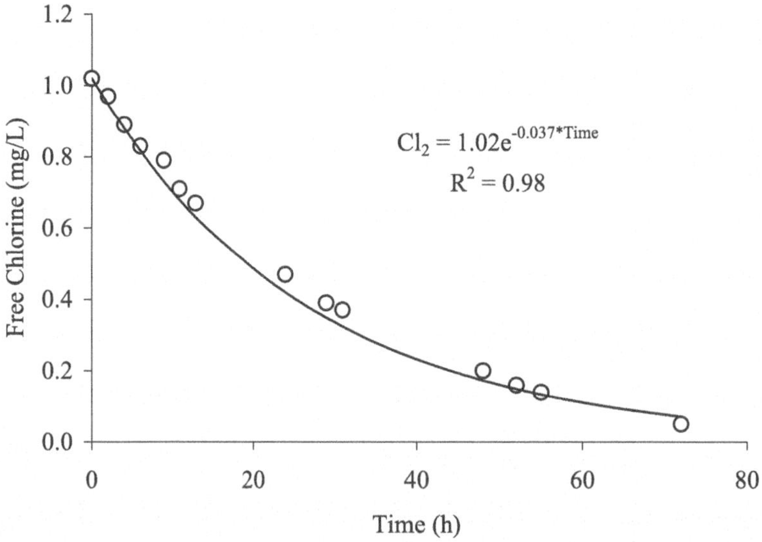

Figure 4.1 Chlorine decay in bulk water taken from OWASA distribution system on 1/15/2001; glass bottle test (pH = 7.54)

Table 4.1
Summary of chlorine decay tests using water from OWASA distribution system

Date of test	Test duration (h)	Initial chlorine concentration (mg/L)	Water temperature (°C)	pH	k_b (h^{-1})
1/8/2001	16	1.03	21	7.6	0.031
1/9/2001	16	1.28	21	7.5	0.031
1/10/2001	17	1.25	21	7.7	0.026
1/11/2001	16.5	1.26	21	7.3	0.038
1/15/2001	72	1.02	21	7.5	0.037
		Average chlorine decay coefficient for bulk water =			0.033

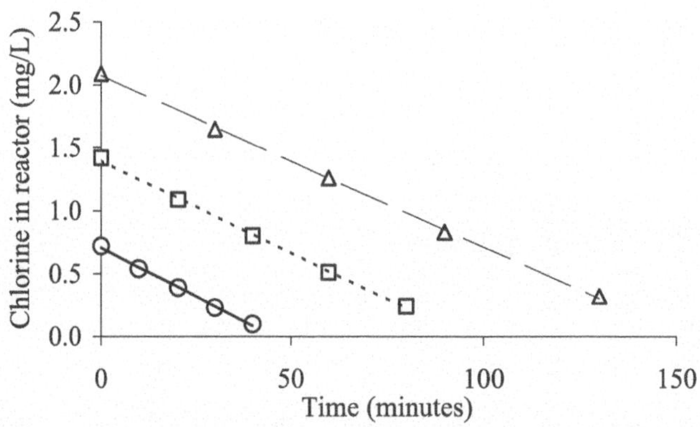

Figure 4.2 Chlorine decay in the CI pipe-section reactor with simulated water velocity = 0.62 m/s, pH = 8.17, and corrosion rate = 3.5 mpy

reaction can be seen by comparing the time needed to for residual chlorine to fall below the detection limit of 0.1 mg/L in the CI pipe section compared to the time needed in the bulk water (Figure 4.1): about 150 min was needed at the highest initial chlorine concentrations in the CI pipe section whereas 60 h was needed in the bulk water.

A zero-order chlorine decay rate indicates that the rate is controlled by the formation of corrosion products, i.e., reductants such as Fe^{2+} rather than by the concentration of chlorine (Vasconcelos et al. 1997). In effect, chlorine reacts very fast with these reductants once they appear. This would be true for corrosion-induced reactions.

A previous study (Kiene, Lu, and Levi 1998) indicated that more than half of the chlorine decay in gray cast iron pipe was due to pipe material interactions. An even higher fractional contribution of wall reaction was found in these experiments, as estimated by:

$$F = \frac{\Delta C_{t,\,w+b} - \Delta C_{t,\,b}}{\Delta C_{t,\,w+b}} \tag{4.3}$$

where

F = fractional decrease in chlorine attributed to wall reaction

$\Delta C_{t,w+b}$ = chlorine consumed after time, t, in the pipe-section reactor due to both wall and bulk decay

$\Delta C_{t,b}$ = chlorine consumed after time, t, due to bulk decay

As an example, the chlorine consumed within the pipe-section reactor (Figure 4.2) after 130 min for the highest initial concentration (2.09 mg/L) is 1.77 mg/L. Using the first-order bulk decay coefficient of 0.033 h^{-1} (from Table 4.1), the corresponding chlorine consumed by the bulk reaction would be only 0.14 mg/L. This is the maximum possible value given that chlorine concentration in the bulk water would be lower due to the wall reaction and thus, the bulk reaction could be even slower. Under these conditions, the resulting F value from Equation 4.3 is 0.92, meaning that 92 percent of the loss in chlorine can be attributed to wall reactions. This high fraction may be partly caused by the loss of the original surface deposit and biofilm due to drying out of the pipe section in storage and failure to regain this surface upon rewetting. The original surface layer and biofilm materials may have been less reactive toward chlorine than the sources of iron corrosion beneath.

The K_w values obtained at each of three water velocities are shown in Figure 4.3. Repeating these experiments at three different initial chlorine concentrations also gave further confirmation of the zero-order nature of the chlorine decay rate. All the K_w values are in the same range as shown in another study of zero-order decay coefficients for 40-year old, unlined CI pipe as shown in the first entry of Table 2.6 (Vasconcelos et al. 1997). K_w increased by a factor of about five when velocity was increased from 0.16 to 0.62 m/s. A similar increase was noted by Hallam et al. (2002) for nearly the same range of velocities. These investigators, however, did not use a laboratory-scale, pipe-section reactor. They used the in-situ field method in which the bulk chlorine decay rate was determined in a batch experiment, chlorine residual was measured at two locations along a pipe, and the wall-decay coefficient was calculated to explain the loss in residual between the two locations that was not accounted by bulk decay; they also estimated water velocity within the pipe. A recent field study was designed to vary the water velocity in a long pipe section and measure K_w (Doshi et al. 2003). While K_w did not correlate with Hazen-Williams

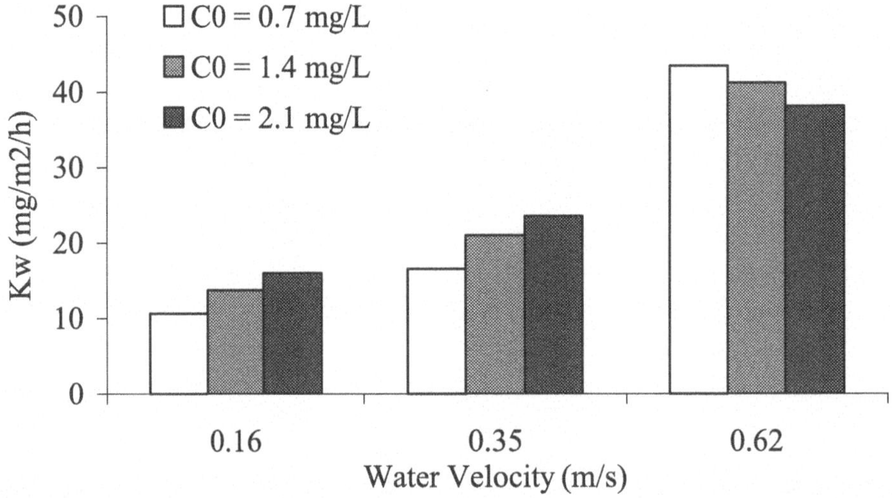

Figure 4.3 Comparison of chlorine decay rate at different water velocities and initial chlorine concentrations in CI pipe-section reactor

C-factor, as had been suggested by others (Vasconcelos et al. 1997), a positive correlation was observed with velocity.

One possible explanation for an increase in K_w with velocity is the corrosion theory put forth by Sarin et al. (2000). These researchers postulate that an increase in velocity can damage the soft scale formed by corrosion within the hard outer shell on the pipe surface. If the hard shell (crust) is broken through, the reservoir of Fe^{2+} in soft inner shell can then release a large amount of Fe^{2+}, which could rapidly react with chlorine. This hypothesis could have implications for design of flushing programs.

Further confirmation of the a zero-order rate model is provided in Figure 4.4 where data are provided at pH of 6.6 and 3.6 in addition to the pH of 8.6 that was used in the initial set of experiments. The rate of chlorine decay as given by the slope of these plots increases with decreasing pH. Two explanations for the variation in wall reaction rate with pH are possible. First, decreasing the pH from the ambient value in this case would convert OCl^- to HOCl, which is more reactive toward Fe^{2+}. Second, the corrosion rate should increase with decreasing pH because H^+ is a major acceptor of electrons released by Fe during the corrosion process ($Fe^0 \rightarrow Fe^{2+}$+2e) (Schock 1999). Rigorous interpretation of the relationship of corrosion rate to pH is not possible because pH increased over the duration of the experiments (approximately two hours). However, the corrosion rate was much higher at the end of the experiment that started at pH of 3.6 than at the end of the experiment that started at ambient pH of 8.6 (7 mpy vs. 4.6 mpy).

A series of four experiments with chloride ion (Cl^-) concentrations ranging from 20 to 160 mg/L over the baseline level in the pipe-section reactor demonstrated again that the chlorine decay rate was zero-order. The K_w values and corresponding corrosion rates are shown in Figure 4.5. The large increase in corrosion rate with increasing Cl^- concentration can be due to its ability to react with dissolved forms of metals that prevent formation of protective chemical precipitates at the pipe surface, to increase the conductivity of the water, and to break down passivating films (Schock 1999, LeChevallier et al. 1993). The fact that the chlorine decay rate did not increase with increasing Cl^- concentration would indicate that the Fe^{2+} release rate had not increased correspondingly. Although the Corrater® probe registered an increase in the corrosion

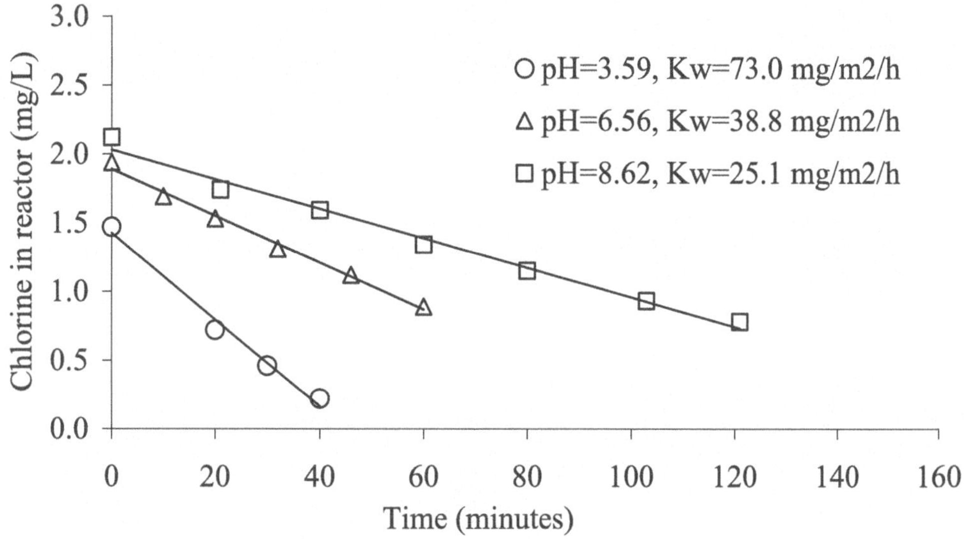

Figure 4.4 Effect of pH on chlorine decay rate in CI pipe-section reactor; simulated water velocity in the pipe-section reactor = 0.16 m/s

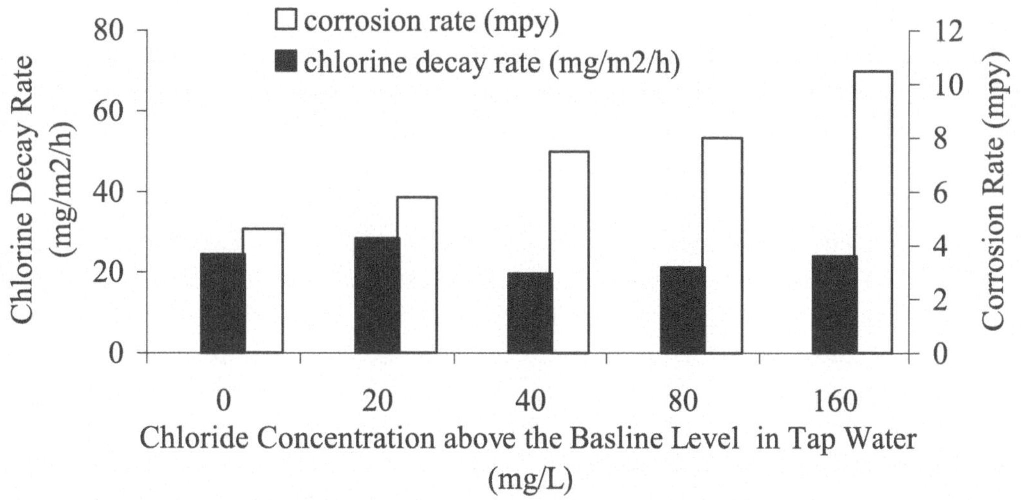

Figure 4.5 Effect of Cl^- concentration on chlorine decay rate and corrosion rate in CI pipe-section reactor; pH = 8.2-8.3

rate with increasing Cl^- concentrations, this probe does not measure the release rate of Fe^{2+} but rather the corrosivity of the water, which is increased by the presence of additional Cl^-. The release rate of Fe^{2+} may have eventually increased but not in the time span of two hours in this experiment.

As indicated in Figure 4.6, the decrease in DO concentration was very slow in comparison to the decrease in chlorine concentration. Whereas nearly all chlorine was lost in less than 3 hours (Figure 4.2), less than 1 mg/L of DO was lost from initial concentrations of 7.3 to 8.6 mg/L in this same time period, regardless of the water velocity. At each simulated water velocity within the reactor, the rate of DO loss appeared to be zero-order. The consumption of DO is most likely due to oxidation reactions involving the pipe wall or the further oxidation of dissolved iron species

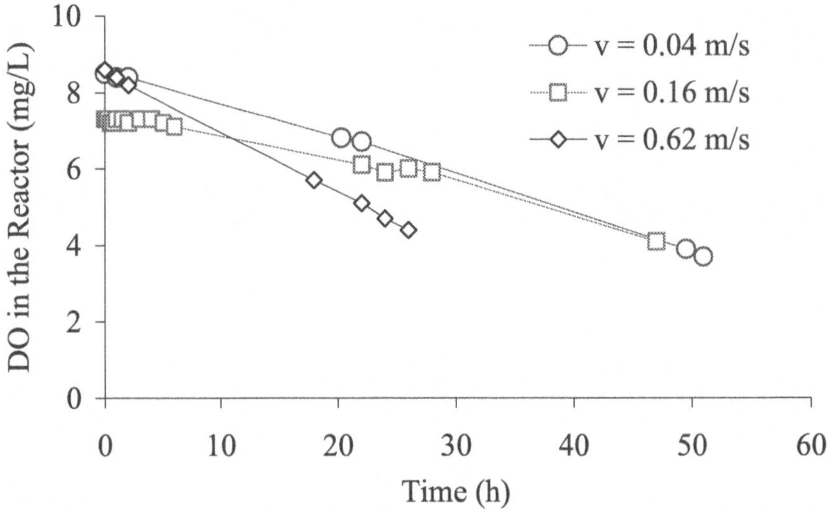

Figure 4.6 Depletion of DO in CI pipe-section reactor at different simulated water velocities (pH = 8.2-8.3 and corrosion rate = 3.6-3.7 mpy)

that are released during corrosion. Although care was taken to eliminate headspace within the reactor, there is a slight possibility for mass transfer of oxygen from air bubbles to the bulk water. This may explain why there was little decrease in DO during the first several hours in the test with a velocity of 0.16 m/s. However, the fact that the largest decrease in DO was found for the highest simulated water velocity (highest stirring speed) does not support oxygen transfer from air to water because this condition would have maximized the oxygen transfer rate if a large headspace was present.

Stagnation experiments were conducted to explore the role of oxygen in control of the release of Fe^{2+} and Fe^{3+} from the pipe wall as discussed recently by Sarin et al. (2000). If stagnation leads to depletion of oxygen, then iron corrosion products can diffuse through the scale layer (the diffusion interface) on the pipe surface into bulk water. The result is a large supply of Fe^{2+} that would increase the rate of chlorine decay. Under flow conditions, on the other hand, the oxygen level is kept high and so, the rate of diffusion is greatly reduced because Fe^{3+} forms a dense shell-like, protective layer around the Fe^{2+}.

Two different experimental designs were used to simulate a low DO condition in the CI pipe-section reactor at a simulated water velocity of 0.16 m/s. The resulting chlorine decay rates are given in Figure 4.7 and these are compared to a typical result for the high DO condition. The chlorine decay rate was again shown in these experiments to be zero-order. When the DO was depleted by recirculation of the water within the pipe-section reactor for three days before addition of chlorine, the chlorine decay rate was over three times faster than that under the usual high DO condition. This would support the idea that protective layer of Fe^{3+} had been lost and that the faster rate of Fe^{2+} release led to a faster rate of chlorine decay. When the DO was depleted quickly by nitrogen gas stripping outside of the reactor instead of more slowly by recirculation of water for three days, the chlorine decay rate was again faster than for the baseline DO condition. However, the increase in chlorine decay rate was not as great (1.4 times faster compared to three times faster for 3-day recirculation experiment).

Although little data are available to confirm this theory, treatment plants that include ozonation may produce a finished water with a much higher dissolved oxygen concentration

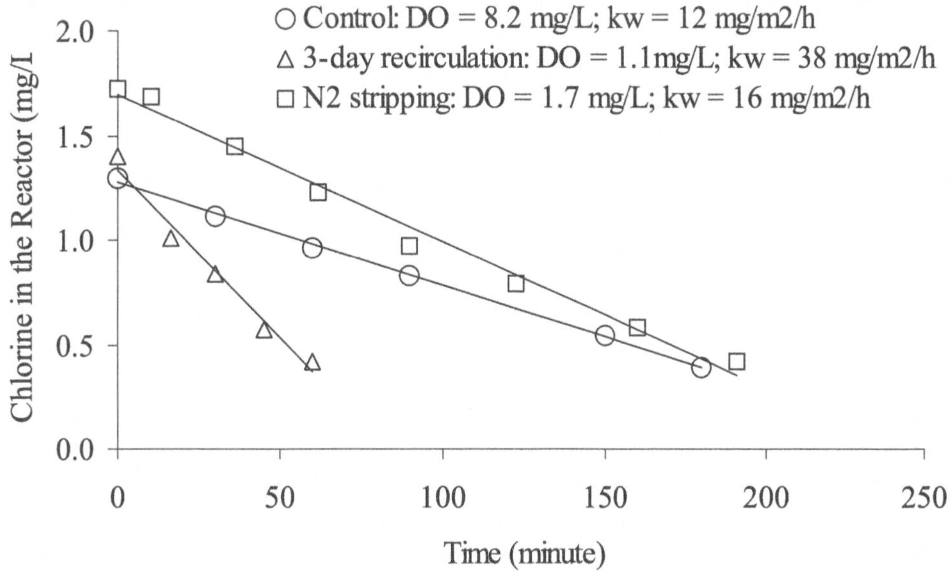

Figure 4.7 Comparison of chlorine decay rates in CI pipe-section reactor under different DO conditions with simulated water velocity = 0.16 m/s, pH = 8.2-8.3, and corrosion rate = 3.5-3.7 mpy

entering the distribution system. Given the findings above, higher oxygen concentrations may reduce the loss of chlorine from the bulk water by preventing deposit-mediated reactions.

Chlorine Decay in DI Pipe-Section Reactor

The chlorine decay observed in DI pipe-section reactor at different simulated water velocities is depicted in Figure 4.8. Also included for comparison is the chlorine decay curve observed in a glass-bottle test. The decay rate, which would represent the sum of bulk water and wall reactions, was much slower in the DI pipe section than that in the CI pipe section. This can be seen qualitatively by comparing the time scales in Figures 4.2 and 4.8.

The data provided in Figure 4.8 also indicate that the overall chlorine decay rate caused by both bulk and wall reaction is not zero-order as was observed for CI pipe in Figure 4.2. A first-order rate process is a much better descriptor of chlorine decay in the DI pipe. The resulting first-order rate coefficient increased significantly with water velocity. This would indicate that a surface reaction was in part responsible for chlorine decay. Vasconcelos et al. (1997) noted that chlorine decay due to wall reactions might be either first- or zero-order. Hallam et al. (2002) designed a pipe-section reactor but without close control of water velocity; they also interpreted the chlorine decay as first-order. A first-order reaction of chlorine with chemical species on the surface of DI pipe may be neither instantaneous nor limited by the rate of production of oxidizable reactants on the surface, as is true for the zero-order reaction on CI pipe. Instead, the reaction rate on DI pipe may be controlled by the concentration of chlorine.

Given that both the bulk water tests in glass bottles and the DI pipe-section reactor tests were described by first-order rate processes, the wall reaction rate coefficient (k_w) can be found assuming that both bulk and wall reactions occur in parallel:

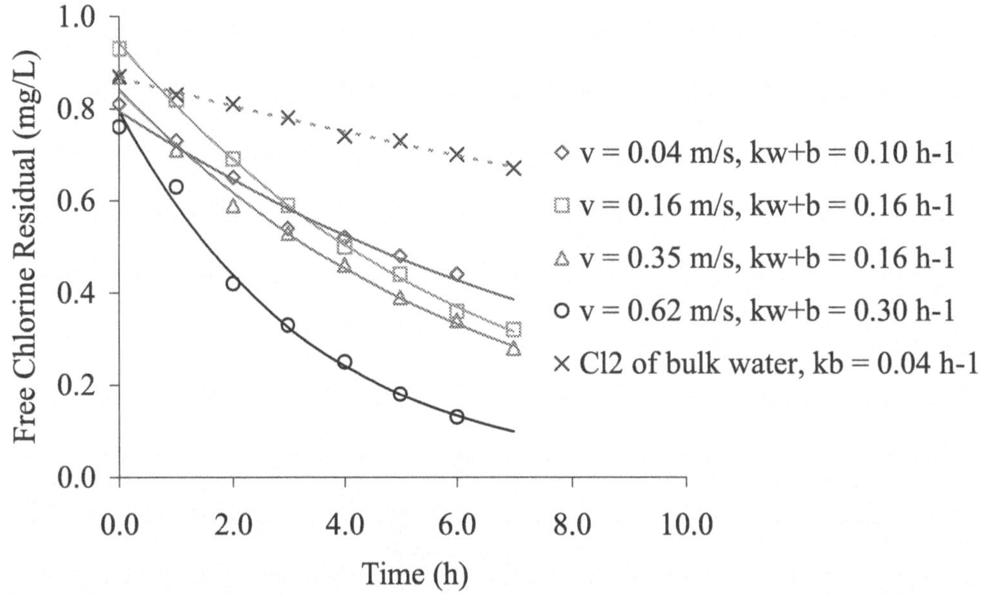

Figure 4.8 Comparison of chlorine decay in DI pipe-section reactor at different simulated water velocities (pH = 7.9, corrosion rate = 3.5 mpy, bulk decay rate in glass bottle also shown)

$$k'_w = k_{w+b} - k_b \tag{4.4}$$

where

k'_w = wall reaction rate coefficient defined as (k_w/r_h) (h^{-1})

k_{w+b} = rate coefficient from DI pipe reactor test (h^{-1})

The fraction of the total decay rate attributable to wall reaction is:

$$F = \frac{k_{w+b} - k_b}{k_{w+b}} \tag{4.5}$$

The calculation based on the DI pipe section test gives the k'_w values that range from 0.07 to 0.26 h^{-1} and the corresponding F values that range from 0.65 – 0.88.

To compare the influence of different water quality parameters on chlorine decay in DI pipe, the k'_w and k_b values resulting from experiments with the DI pipe-section reactor for tap water, tap water amended with 20 mg/L of Cl^-, and tap water amended with HCl to produce a pH of 6.1, are given in Figure 4.9. Neither k_w nor k_b was changed significantly by either amendment. The lack of a response to changes in water quality in this case is thought to be primarily because the cement lining on the DI pipe prevents the release of Fe^{2+} regardless of increases in the Cl^- or decreases in the pH. For comparison, the increase in the zero-order rate coefficient, K_w, for the CI pipe upon a similar decrease in pH was about a factor of three (see Figure 4.4).

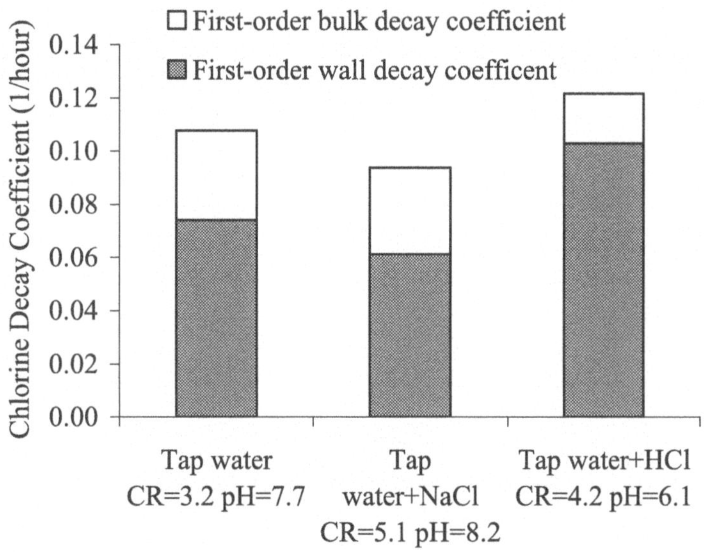

Figure 4.9 Comparison of decay coefficient at different pH, with and without addition of 20 mg/L NaCl as Cl^- (CR = corrosion rate (mpy)

Table 4.2
Raw and finished water quality at North Chatham County WTP (October 26, 2000)

Parameter	Raw	Finished
Turbidity (ntu)	5.9	0.13
Fe (mg/L)	0.18	0.01
Mn (mg/L)	0.288	0.001
pH	7.16	7.28

FIELD MEASUREMENTS IN NORTH CHATHAM COUNTY, N.C.

General Water Quality Indicators

Raw and finished water quality parameters for North Chatham County system on October 26, 2000, the day of the first sampling in the DS are given in Table 4.2. These data show that the WTP was removing iron and manganese and producing a water of generally acceptable quality. At each sampling location, measurements were made of free chlorine residual, pH, temperature, DO, and corrosion rate. The results obtained on October 26, 2000 and November 22, 2000 are summarized in Tables 4.3 and 4.4, respectively. The results from the first sampling date included two sites (Locations 3 and 4) with negligible chlorine residual (see Figure 3.3 for sampling locations). In addition, a sulfurous odor was detected at the dead-end sample site (Location 4). These sites may have problems in addition to corrosion that would confound the corrosion measurement results. Therefore, these sites were replaced by Location 5, which is mid-way between locations 2 and 3, for sampling on November 22, 2000. Measurement of ferrous and total iron concentrations were added as shown in Table 4.4.

Table 4.3
Sampling results in North Chatham County DS on October 26, 2000

Location	Free chlorine (mg/L)	pH	Temperature (°C)	DO (mg/L)	Corrosion rate (mpy)
1	2.36	7.28	19.5	6.9	5.1
2	1.72	7.76	19.0	6.5	4.6
3	0.04	8.07	22.7	6.0	4.7
4	0.00	9.60	21.0	2.2	5.0

Table 4.4
Sampling results in North Chatham County DS on November 22, 2000

Location	Free chlorine (mg/L)	pH	Temp. (°C)	DO (mg/L)	Corrosion rate (mpy)	Ferrous iron (mg/L)	Total iron (mg/L)
1	2.38	7.26	13.0	7.3	4.3	0.00	0.01
2	1.97	-	14.7	7.4	4.3	0.02	0.05
5	1.57	-	14.0	7.1	4.2	0.01	0.05

The manufacturer of the Corrater probe specifies that the measurement error is ±0.5 mpy. Therefore, the differences in measured corrosion rates within each sampling data set cannot be considered to be significant. However, the corrosion rate for finished water was significantly lower on November 22 compared to October 26. This may be attributed to the decrease in temperature from 20°C to 13°C.

Bulk Chlorine Decay Rate

Bulk chlorine decay tests were conducted on grab samples returned to the laboratory on each of the two sampling dates. The results are presented in Figures 4.10 and 4.11, respectively. All the data fit a first-order decay model rather well. The k_b value was about the same for the two samples of finished water leaving the WTP as may be expected for a consistently treated lake water source over a one month interval of sampling. However, k_b within the DS was markedly greater than in the finished water leaving the WTP on each date. If the bulk decay rate was truly first-order with respect to chlorine concentration, the rate coefficient should not change with location in the DS even though the concentration of chlorine at each location may be different. The same trend of increasing k_b within the DS was observed in previous studies of chlorine decay conducted Durham and Raleigh, N.C. (DiGiano et al. 2000, 2001).

The bulk decay rate could depend upon both the concentration of chlorine and the concentration of chlorine-oxidizable substances. This concept has been put forth by DiGiano et al. (2000) and by Clark and Sivaganesan (1998) as an explanation for failure of first-order decay models. The reaction scheme can be written in general terms as:

$$Cl_2 + A \rightarrow \text{Oxidized Products} \qquad (4.6)$$

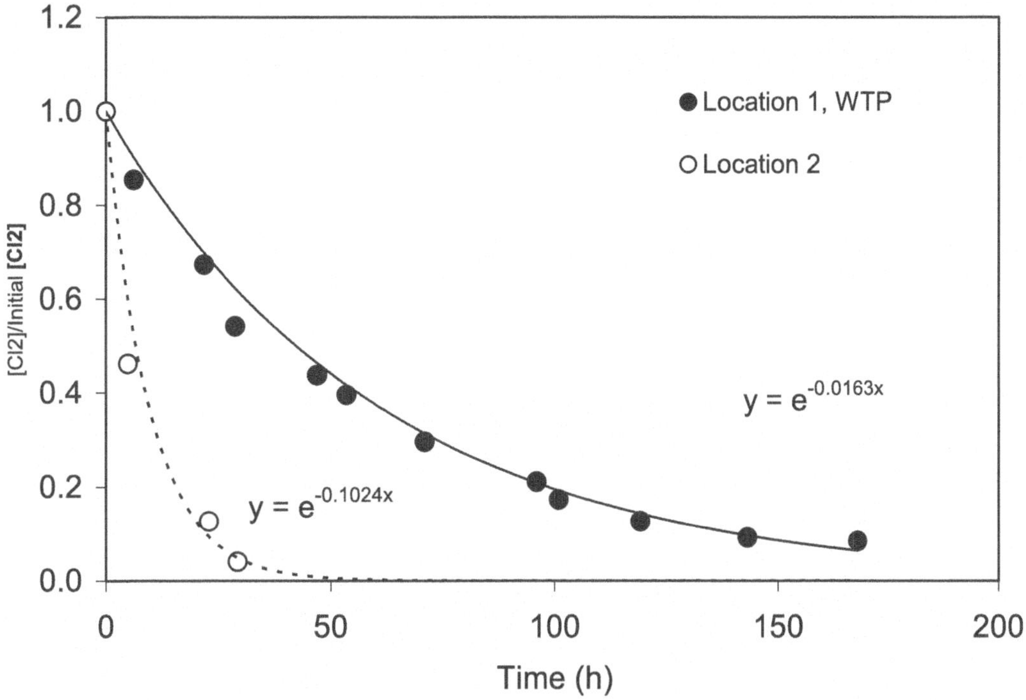

Figure 4.10 Bulk chlorine decay rate on October 26, 2000 in North Chatham County at room temperature (22°C)

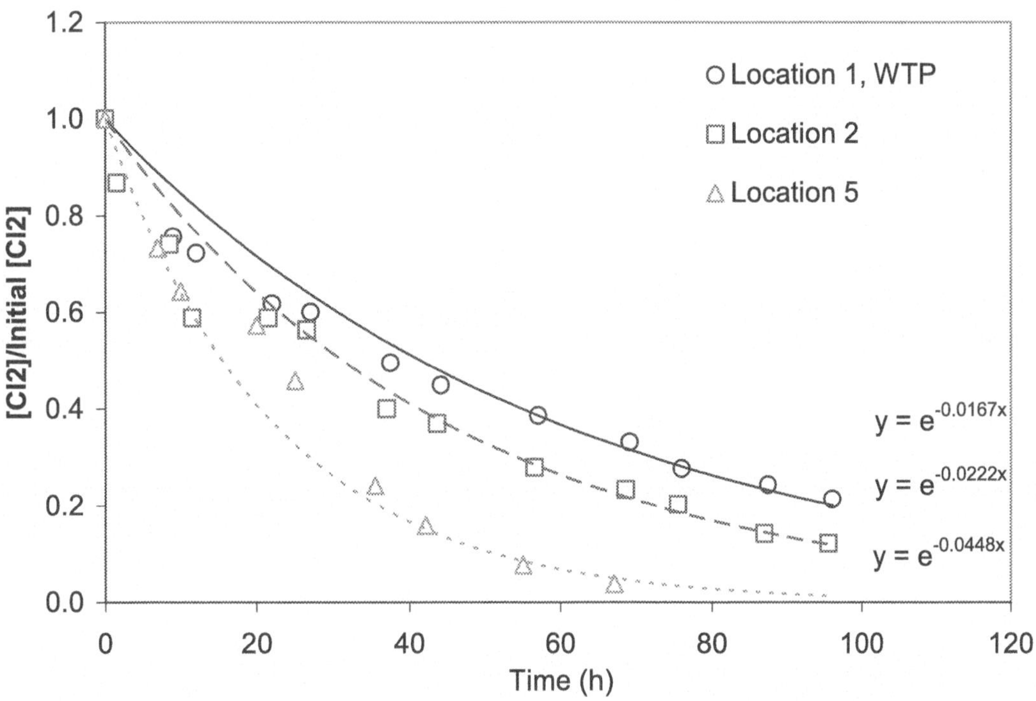

Figure 4.11 Bulk chlorine decay rate on November 22, 2000 in North Chatham County at room temperature (22°C)

where

A = total concentration of chlorine-oxidizable substances

If the initial chlorine concentration $[Cl_2]_o$ is far in excess of A, the reaction could be considered pseudo first-order with respect to A such that:

$$\frac{d[Cl_2]}{dt} = -k_{ps}[A] \tag{4.7}$$

where

$k_{ps} = k[Cl_2]_o$ = pseudo first-order rate coefficient

The chlorine concentration must still decrease in time, but the impact of the change in its concentration on reaction rate is small compared to that of the change in concentration of A.

The above argument can be carried further. The residual concentration of A, which is not measurable, can be related to the residual concentration of chlorine by assuming a reaction with stoichiometry of 1:1 so that:

$$[A] = [A]_o - \{[Cl_2]_o - [Cl_2]\} \tag{4.8}$$

where

$[A]_o$ = initial concentration of chlorine-oxidizable substances

Substitution of Equation 4.7 into Equation 4.6 and integration with the initial condition given by t=0 and $[Cl_2] = [Cl_2]_o$ yields:

$$\ln\frac{[Cl_2] - [Cl_2]_o + [A]_o}{[A]_o} = -k_{ps}t \tag{4.9}$$

Stoichiometry requires that the total chlorine demand ($[Cl_2]_o$ - $[Cl_2]$) approach $[A]_o$ after a long time period. If $[Cl_2]_o$ is held constant such that k_{ps} is constant, Equation 4.9 can be used to generate $[Cl_2]$ vs. t data for different values of $[A]_o$, all being much smaller than $[Cl_2]_o$. It can be shown that the resulting data will closely approximate a straight line if plotted according to the first-order rate model in chlorine concentration provided that $[Cl_2]_o >> [A]_o$. Thus, chlorine decay kinetics could be incorrectly interpreted as a first-order dependency on chlorine concentration. Furthermore, the slopes of such plots will increase as the value of $[A]_o$ increases. If the concentration of chlorine oxidizable substances in*creases at different locations within the DS due to release of corrosion products, the observed* k_b would also increase, as was noted in Figures 4.10 and 4.11.

Chlorine Wall Reaction Coefficient

The chlorine wall reaction component of the total chlorine decay can be calculated for the section of pipe between Locations 1 and 2. This pipe is 16 in. (40.6 cm) in diameter and 5.3 mi

(8.5 km) in length. The flow rate in this pipe was 0.95 mgd (3.6×10^3 m^3/d) on October 26, 2000 and 0.92 mgd (3.5×10^3 m^3/d) on November 22, 2000. No other pipes are connected to this line and thus, the time of travel between Locations 1 and 2 is about 450 min based on the calculated velocity of about 1 ft/s (0.32 m/s) and the pipe length. Assuming a first-order reaction, chlorine decay in a pipe can be described by Equation 2.16

$$\frac{C_2}{C_1} = e^{-(k_b + k_w')t} \tag{2.16}$$

such that:

$$C_2 = C_1 \exp(-k_{tot}t) \tag{4.10}$$

where

C_1 = chlorine concentration (mg/L) at Location 1
C_2= chlorine concentration (mg/L) at Location 2
t = travel time between Locations 1 and 2
$k_{tot} = k_b + k'_w$
k'_w = wall decay rate constant in units of time^{-1}

The wall reaction rate coefficient can then be expressed as k_w in units of length/time by multiplying by the hydraulic radius of the pipe, r_h. The resulting values of k_w are summarized in Table 4.5 along with measured bulk decay rate coefficients and corrosion rates. It should be noted that the k_b values were first adjusted for the field temperatures using the Arrhenius equation and the commonly assumed activation energy (Weber and DiGiano 1996) which gives the following expression:

$$k_{b,\,field} = k_{b,\,lab}(1.047)^{(T_{field} - T_{lab})} \tag{4.11}$$

The wall reaction coefficients found for North Chatham County are within the general range of literature values as listed in Table 2.4. The calculated wall reaction coefficient was 40 percent lower on the second sampling date than the first, showing that temperature plays a significant role in chlorine decay related to corrosion. The corrosion rate measurements were 15 percent lower in the second sample. Thus the trends in corrosion rate measurement seem to follow the trends in wall reaction over time. The lack of precision in the corrosion rate measurements (margin of error of 0.5 mpy) using this particular instrument in the relatively mildy corrosive environment of the DS makes it difficult to define a predictive relationship between corrosion rate and wall reaction, particularly when trying to compare measurements made at different locations in the DS on the same day. However, corrosion rate measurements do provide a good indicator of a shift in the aggressiveness of the water which may translate to a corresponding change in wall reaction reaction rates for disinfectants.

Table 4.5
Chlorine decay coefficients and corrosion rates for North Chatham County

Sampling date	Field temperature (°C)	k_b* (d^{-1})	k'_w (d^{-1})	k_w (ft/d)	Corrosion rate† (mpy)
10/26/00	20	0.36	0.154	0.463	5.1 ± 0.5
11/22/00	13	0.26	0.095	0.285	4.3 ± 0.5

*Adjusted for field temperature from laboratory rate data.
†Corrosion rate in finished water.

Table 4.6
Finished water quality in 2001 for Durham, N.C.

Parameter	Value
Alkalinity (mg/L as $CaCO_3$)	24
pH (range)	7.0-7.2
Conductivity (μohms/cm)	142
Hardness (mg/L as $CaCO_3$)	41
Chloride (mg/L)	10.2
TOC (range) (mg/L)	2.2-4.9
Turbidity (range) (ntu)	0.10-0.15
Orthophosphate (mg/L P)	0.27

Data from: <http://www.ci.durham.nc.us/departments/environ/water_quality_report_01.pdf>

FIELD MEASUREMENTS IN DURHAM, N.C.

Finished Water Quality

The water quality characteristics for finished water in Durham, N.C. are presented in Table 4.6 as annual averages (except for pH, TOC and turbidity which are given as ranges) in 2001. As is typical of this region, finished waters are very low in alkalinity and hardness and thus corrosion control is important.

Bulk Chlorine Decay Rate

The results of bulk chlorine decay tests on grab samples of finished water from the Brown and Williams WTP are shown in Figure 4.12. A first-order decay model was fit to the data to give k_b of 0.0165 h^{-1} (R^2 of 0.97) for the Williams WTP and 0.0155 h^{-1} (R^2 of 0.97) for the Brown WTP.

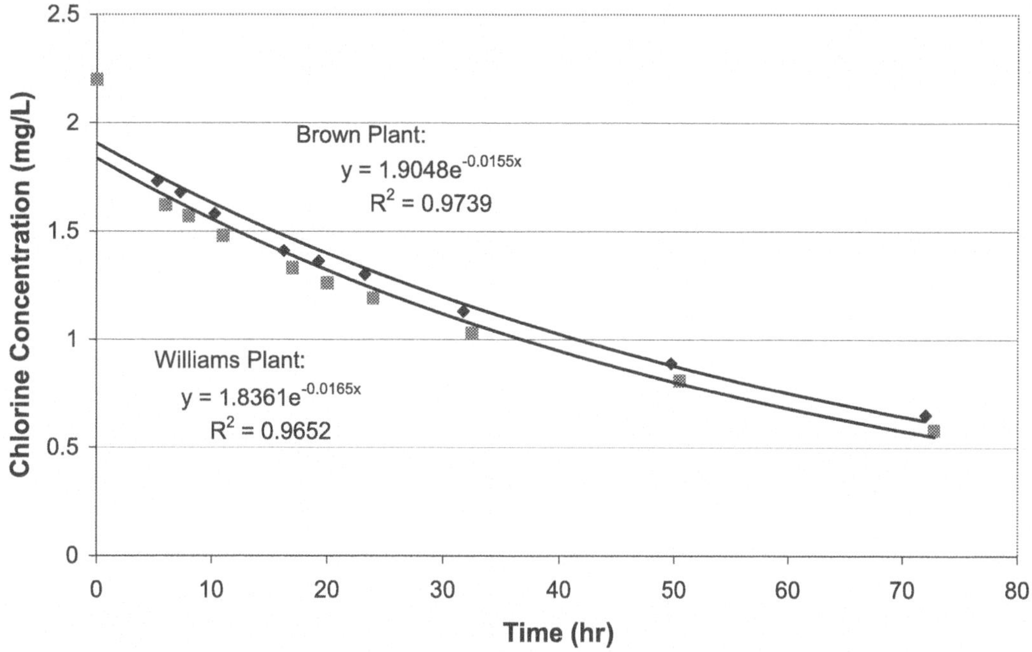

Figure 4.12 Bulk chlorine decay coefficients for Durham, N.C. at a temperature of 28°C

Chlorine Wall Reaction Coefficient

Using the hydraulic base model in EPANET calibrated by Pitometer Associates, a water quality model was created to predict chlorine concentrations in the distribution system. Grab samples were collected during a 5-day period in July, 2001 and analyzed for free chlorine concentration in the field. The average flow during the sampling period was 33.22 mgd (125.6×10^3 m^3/d), slightly higher than the average day demand in 2001. During the five-day sampling period in July 2001, the City of Durham had to shut down the Brown WTP for several hours; this WTP usually provides about two-thirds of the total production. Although the shutdown occurred during the night, the elevated storage tank levels dropped significantly. Therefore the sampling period contained a higher degree of water turnover in the tanks than is normally seen in daily operations.

Once a hydraulic base model was established to replicate flows during the sampling period, the bulk chlorine residual component was added to start the calibration process. Because EPANET only allows a single value of the bulk decay coefficient to be globally specified, the average of the two WTP bulk decay coefficients, 0.0160 h^{-1} or 0.384 day^{-1}, was used in the model. The initial values of chlorine concentrations at nodes in the calibration model were set by geographic location in accordance with general trends in the samples collected. The finished water leaving both WTPs had an average free chlorine concentration of 2.1 mg/L during the sampling period. The areas near the WTPs served by large diameter transmission mains were initialized at 1.50 mg/L. The majority of the service area was initialized to 1.0 mg/L, with outlying areas set to 0.5 mg/L.

Little data were available on chlorine concentrations in the tanks, as the City of Durham does not routinely sample them. Based on information provided by the operators, the initial

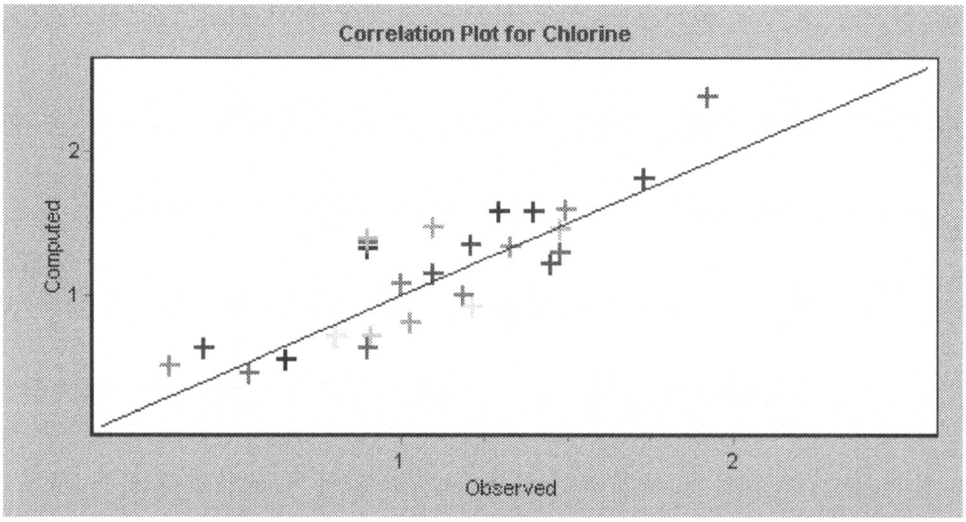

Figure 4.13 Calibration of EPANET model for free chlorine residual in DS of Durham, N.C.

chlorine concentrations were estimated for the four storage tanks. The tanks were considered to be completely mixed for modeling purposes.

The calibration of the EPANET model for free chlorine residual incorporated a first-order wall reaction using Equation 2.18 to correlate k_w to the Hazen-Williams C-factor with an α value of 25. The resulting k_w was 0.543 ft/day for old CI pipe with a C-factor of 46. For newer DI pipe with a C-factor of 130, this resulted in a k_w of 0.192 ft/day. These results correspond to wall reaction coefficients found in the literature, which generally range from 0 to 5 ft/day (Vasconcelos et al. 1997).

The tank reaction rate coefficient was set to 0.6 day^{-1}, which is approximately 1.5 times greater than the bulk chlorine reaction rate. The observed and predicted values of free chlorine residual results with this α value are compared in Figure 4.13. The correlation coefficient for this calibration was 0.845 with four outliers removed from the data set. The outliers included dead-end sample locations where water age in the 72-hour model run was insufficient to simulate the low measured residuals and other samples where historical chlorine concentrations did not fit with model predictions.

While the global calibration factor provided a fairly good overall result, the wall reaction coefficients were adjusted individually in several locations to provide a better match to sample data. A summary of the range of k_w values obtained for different pipe material is given in Table 4.7.

The maximum values of k_w for the relatively new DI pipe should be not be considered as representative. The higher values (2.0 ft/day) were assigned to individual pipe segments in the effort to calibrate the model in areas near sample locations. Using the current calibration protocols, the wall reaction coefficient serves as the adjustment in the model for all sources of error. Therefore, high wall reaction coefficients may reflect other sources of modeling error, such as misclassification of pipe diameter or material, existence of unknown valves, or water quality factors including extraneous sources of chlorine demanding substances in the bulk water. The corrosion rate as measured by the AquaMate Portable Corrater® does not appear to give a strong indicator of k_w. The range of k_w values for old DI pipe in the Durham DS is similar to those found by in-situ measurements on a DI line in the North Chatham County DS (Table 4.5) while the

Table 4.7
Wall reaction coefficients for different pipe materials/conditions in the DS of Durham, N.C. after calibration of EPANET model

Pipe C-factor	Type of pipe	k_w (ft/day)	
		min	max
≤100	CI	0.25	1.32
100 - 120	Old DI	0.21	0.25
≥120	Relatively new DI	0.17	2.0

corrosion rate in the Durham system was higher than in the North Chatham County system (6.1 - 6.4 mpy at the two WTPs in Durham vs. 4.3 to 5.1 mpy within the North Chatham County system as shown in Table 4.5). The finished water quality characteristics in both systems are general similar if expressed in terms of alkalinity, pH and TOC. The higher corrosion rate in the Durham system may be due to higher temperature given that the Durham system samples were collected in mid-summer (28°C) and the North Chatham County system samples were collected in mid-fall (20°C and 13°C on two sampling dates).

CHAPTER 5
CONCLUSIONS AND RECOMMENDATIONS

The pipe-section reactor is a convenient bench-scale device to study the effects of water velocity, water quality parameters and pipe materials on corrosion and chlorine decay rate. For the old, heavily tuberculated CI pipe, a zero-order kinetic model described the overall chlorine decay rate quite well for a range of initial chlorine concentrations, pH, Cl^- concentration and DO conditions. A zero-order model would be consistent with wall reactions that dominate the total chlorine decay rate and control of the reaction rate by the generation of corrosion products rather than by the concentration of chlorine. Chlorine decay tests in glass bottles confirmed that the decay rate in bulk water was much slower than the overall decay in the pipe-section reactor; moreover, the decay rate in bulk water was described by first-order kinetics with respect to chlorine concentration rather than zero-order kinetics.

Factors that increased the zero-order, wall reaction rate coefficient were higher velocity (due to increased mass transfer at the pipe surface), lower pH (due to faster corrosion rate), and lower DO (due to greater release of Fe^{2+} after removal of the oxidized layer at the surface). Surprisingly, increasing the Cl^- concentration did not immediately increase the zero-order rate coefficient although it greatly increased the corrosion rate.

Unlike the CI pipe section, the chlorine decay kinetics were first-order with respect to chlorine concentration in the new, cement-lined DI pipe section. While the wall reactions were still important contributors to chlorine decay, the rate of decay was about five times slower in the DI pipe section than in the CI pipe section. The shift to first-order decay kinetics could suggest that chlorine becomes the limiting reactant in DI pipe.

Field measurements of corrosion rate and disinfectant decay were obtained in two different distribution systems within about 40 miles (84 km) of one another. The raw water supplies are surface waters of low alkalinity and hardness, with less than 30 mg/L as $CaCO_3$, as is typical of the Piedmont region of N.C. The raw waters are treated by alum coagulation, sedimentation and filtration. At the time of this study, free-chlorine was used in Durham and North Chatham County. While general statements about factors that cause corrosion will always have exceptions (Schock 1999), protective scales usually do not readily form on pipe materials if alkalinity and hardness are low; moreover, low alkalinity also infers little capacity to resist change in pH once corrosion begins. Thus, corrosion control is important at each water utility. Corrosion rates as measured in bulk water samples using the AquaMate Portable Corrater®, ranged from 4.3 to 6.1 mpy. These would be considered low to moderate values based on other field studies (Norton and LeChevallier 1997).

The results from the two systems are difficult to compare directly because different techniques had to be used to investigate the relationship between chlorine wall reaction rate and corrosion rate. The results for each system, however, provide insights into the importance of corrosion reactions in reducing disinfectant residual as well as into the advantages and disadvantages of measurement techniques to quantify these reactions.

Electrochemical measurements of corrosion rate with bulk water samples do not account for on-going surface reactions at pipe walls and thus, such information is of limited value. Weak evidence of relationship between corrosion rate and disinfectant decay rate by wall reactions was provided from the results obtained in North Chatham County DS. A long section of DI pipe was used in the North Chatham County DS to measure free chlorine concentration at two locations.

When coupled with the results for the decay rate in bulk water, the k_w values were calculated for two different grab samples collected about one month apart. The corrosion rates were on these two sampling dates were 4.3 mpy and 5.1 mpy. The lower corrosion rate corresponded to a temperature that was 7°C lower. After accounting for a measurement error in corrosion rate of ±0.5 mpy, a small difference in corrosion rate is suggested, possibly influenced by the temperature difference. The calculated k_w values for corrosion rates of 4.3 mpy and 5.1 mpy were 0.094 d^{-1} and 0.154 d^{-1}, respectively. In this case, the trend of decreasing wall reaction rate constant with temperature is reflected in a decrease in measured corrosion rate. However, it is difficult to define a predictive relationship between corrosion rate and wall reaction rate given the lack of precision in the corrosion rate measurements with the instrument used in this study.

With more than 1,000 miles of pipe, the Durham DS was too large and complex to isolate a section of CI pipe for determinations of k_w by the in-situ method as used in the North Chatham County DS. However, a hydraulically-calibrated EPANET model was available. The model was calibrated using α, a constant of proportionality between k_w and the Hazen-Williams C-Factor, to provide the best fit to free chlorine measurements made throughout the DS. The α value for the entire DS then determines k_w for individual pipe sections based on their C factor. This technique for determining k_w results in a range of values even within one type of pipe material if the C factors are different. The resulting k_w values ranged from 0.17 to 2 ft/d for this distribution system.

Little further interpretation is possible for the range in k_w owing to the uncertainty associated with the calibration method. Fewer than 100 locations scattered throughout the DS were used for free chlorine residual measurements. These provide the only data from which k_w values are assigned to nearly 12,000 pipe segments in the calibration procedure. Obviously, an increased number of samples would decrease the uncertainty in estimates of k_w. While in this case a calibration procedure was used that correlates C-factor to k_w, this relationship has not been proven. Finally, information about the C-factor for each pipe is based on hydraulic calibration which is also subject to uncertainty as well.

The k_w values for old DI pipe for the Durham and North Chatham County distribution systems were similar (0.21 – 0.25 ft/d vs. 0.094 – 0.154 ft/d) although measured by two different methods (in-situ method in North Chatham County vs. EPANET calibration in Durham). The corrosion rates were somewhat different (4.3 mpy – 5.1 mpy in North Chatham County vs. 6.1 - 6.4 mpy in Durham). The data are too limited to draw further conclusions; however, a relationship between corrosion and free chlorine decay is possible.

The field studies did not provide a specific relationship that would provide a method of estimating k_w from measurement of corrosion rate in the bulk water. The results showed the importance of unique characteristics found in every DS that dictate the approach to investigating k_w and the inherent drawbacks to corrosion rate measurements in bulk water that cannot account for surface reactions on pipe walls. To conduct more field studies in order to increase the database on k_w and its relationship to corrosion would be costly and time-consuming and they would probably still not be of general use to all water utilities. Field studies are nevertheless of value for each specific utility albeit these can be costly.

The laboratory studies using the pipe-section reactor hold more promise than field studies in quantifying a relationship among fundamental pipe characteristics (e.g., C-factor, age, material, corrosivity), water quality, water velocity and k_w. The pipe-section reactor studies conducted in this research could be greatly expanded to examine a wide range of pipe materials and age, water velocity and disinfectant type (e.g., chloramines vs. free chlorine). Nevertheless, the need for confirmation of laboratory results in field studies remains important.

REFERENCES

Abernathy, C.G., and A.K. Camper. 1998. The Effect of Phosphorus Based Corrosion Inhibitors and Low Disinfectant Residuals on Distribution Biofilms. In *Proc. of the AWWA Water Quality Technology Conference,* San Diego, Calif.

Appenzeller, B.M.R., M. Batte, L. Mathieu, J.C. Block, V. Lahoussine, J. Cavard, and D. Gatel. 2001. Effect of Adding Phosphate to Drinking Water on Bacterial Growth in Slightly and Highly Corroded Pipes. *Water Research*, 35(4):1100-1105.

Baribeau, H., M. Prévost, R. Desjardins, and P. Lafrance. 2001. Changes in Chlorine and DOX Concentrations in Distribution Systems. *Jour. AWWA*, 93(12):102-114.

Blenke, H. 1978. Mixing and Dispersing in Loop Reactors. Edited by M. Bruxelmane and S. Lefebvre. C7-1-C7-55. In *Proc. International Symposium on Mixing Proceedings*. Mons, France: Faculte Polytechnique de Mons.

Characklis, W.J., and K.C. Marshall. 1990. *Biofilms*. New York: John Wiley and Sons.

2001 City of Durham Water Quality Report. 2001. [Online.] Available: <http://www.ci.durham.nc.us/departments/environ/water_quality_report_01.pdf> [cited August 15, 2003]

Clark, R.M., and M. Sivaganesan. 1998. Predicting Chlorine Residuals and Formation of TTHMs in Drinking Water. *Jour. Environmental Engineering, ASCE*, 124(12):1203-1210.

DiGiano, F.A., D.F. Francisco, W. Zhang, and L. Todd. 2000. Bacterial Regrowth in Drinking Water Distribution System: A Comparison of Durham and Raleigh, N.C. Report No. 50212. Raleigh, N.C.: Water Resources Research Institute of the Univ. of North Carolina.

DiGiano, F.A., W. Zhang, D.F. Francisco, and M. Wood. 2001. Data Collection to Support a Simplified Bacterial Regrowth Model for Distribution Systems. Report No. 331. Raleigh, N.C.: Water Resources Research Institute of the University of North Carolina.

Doshi, B., W.M. Grayman, D. Guastella, and T. Young. 2003. Field Testing the Chlorine Wall Demand in Distribution Systems. In *Proc. of the AWWA Water Annual Conference,* Anneheim, Calif.

Frateur, I., C. Deslouis, L. Kiene, Y. Levi, and B. Tribollet. 1999. Free Chlorine Consumption Induced by Cast Iron Corrosion in Drinking Water Distribution System. *Water Research*, 33(12):1781-1790

Gauthier, V., B. Gerard, J.M. Portal, J.C. Block, and D. Gatel. 1999. Organic Matter as Loose Deposits in a Drinking Water Distribution System. *Water Research*, 33(4):1014-1026.

Hallam, N.B., J.R. West, C.F. Forster, J.C. Powell, and I. Spencer. 2002. The Decay of Chlorine Associated with the Pipe Wall in Water Distribution Systems. *Water Research*, 36:3479-3488.

Hua, F., J.R. West, R.A. Barker, and C.F Forster. 1999. Modelling of Chlorine Decay in Municipal Water Supplies. *Water Research,* 33(12):2735-2746.

Kiene, L., W. Lu, and Y. Levi. 1998. Relative Importance of the Phenomena Responsible for Chlorine Decay in Drinking Water Distribution Systems, *Water Sci. and Technol.*, 38(6):219-227.

Knocke, W.R., J.E. Van Benschoten, M. Kearney, A. Soborski, and D.A. Reckhow. 1990. *Alternative Oxidants for the Removal of Soluble Iron and Manganese.* Denver, Colo.: AwwaRF.

LeChevallier, M.W., C.D. Lowry, and R.G. Lee. 1990. Disinfection of Biofilms in a Model Distribution System. *Jour. AWWA*, 82(7):87-99.

LeChevallier, M.W., C.D. Lowry, R.G. Lee, and D.L. Gibbon. 1993. Examining the Relationship Between Iron Corrosion and the Disinfection of Biofilm Bacteria. *Jour. AWWA*, 85(7):111-123.

LeChevallier, M.W., N.J. Welch, and D.B. Smith. 1996. Full-scale Studies of Factors Related to Coliform Regrowth in Drinking Water. *Appl. and Environ. Microbiol.*, 62(7):2201-2211.

LeCourt, T.D., D.E. Pinsky, J.L. Anderson, and R.J. Young. 1999. Using Water Quality Modeling to Address Distribution System Water Quality Concerns. In *Proc, of the AWWA Water Quality Technology Conference*, San Diego, Calif.

Li, S.Y., P. Biswas, M.C. Meckes, R.M. Clark, M. Dosani, and R. Krishnan. 2000. Chlorine Decay and Biofilm Studies in a Pilot Scale Drinking Water Distribution Dead End Pipe System. In *Proc. of the AWWA Water Quality Technology Conference*, Salt Lake City, Utah.

Maier, S.H., R.S. Powell, and C.A. Woodward. 2000. Calibration and Comparison of Chlorine Decay Models for a Test Water Distribution System. *Water Research*, 34(8):2301-2309.

Norton, C.D., and M.W. LeChevallier. 1997. Chloramination: Its Effect on Distribution System Water Quality. *Jour. AWWA*, 89(7):66-77.

Olson, R.M., and S.J. Wright. 1990. *Essentials of Engineering Fluid Mechanic*. New York: Harper & Row.

Pitometer Associates. 2001. *Hydraulic Model for the City of Durham, NC*. Forthcoming.

Powell, J.C., N.B. Hallam, J.R. West, C.F. Forster, and J. Simms. 2000. Factors Which Control Bulk Chlorine Decay Rates. *Water Research*, 34(1):117-126.

Raad, R., A. Rompre, M. Prevost, C. Morisette, and P. Lafrance. 1998. Selecting the Best Corrosion Control Strategy Using a Rapid Corrosion Monitoring Tool. In *Proc. of the AWWA Water Quality Technology Conference*, San Diego, Calif.

Rossman, L.A. 2000. Computer Models/EPANET. In *Water Distribution Systems Handbook*. Edited by L. Mays. New York: McGraw-Hill.

Rossman, L.A., R.A. Brown, P.C. Singer, and J.R. Nuckols. 2001. DBP Formation Kinetics in a Simulated Distribution System. *Water Research*, 35(14):3483-3489.

Rossman, L.A., R.M. Clark, and W.M. Grayman. 1994. Modeling Chlorine Residuals in Drinking Water Distribution Systems. *Jour. Environ. Engrg., ASCE*, 120(4):803-821.

Sarin, P., J. Bebee, M.A. Beckett, K.K. Jim, D.A. Lytle, J.A. Clement, W.M. Kriven, and V.L. Snoeyink. 2000. Mechanism of Release of Iron from Corroded Iron/Steel Pipes in Water Distribution Systems. In *Proc. of the AWWA Annual Conference,* Chicago, Ill.

Sarin, P., V.L. Snoeyink, J. Bebee, W.M. Kriven, and J.A. Clement. 2001. Physico-Chemical Characteristics of Corrosion Scales in Old Iron Pipes. *Water Research,* 35(12):2961-2969.

Schock, M.R. 1999. Internal Corrosion and Deposition Control. In *Water Quality and Treatment*. New York: McGraw-Hill.

Smith, A.E. 1994. Estimated Compliance of North Carolina Water Treatment Plants with Proposed Enhanced Coagulation Requirements. MS report. University of North Carolina, Chapel Hill, N.C.

Snoeyink, V.L., and I. Wagner. 1996. Principles of Metallic Corrosion in Water Distribution Systems. In *Internal Corrosion of Water Distribution Systems*, 2nd ed. Denver, Colo.: AwwaRF/DVGW-TZW.

Valentine, R.L., P.J. Vikesland, B.D. Angerman, S.A. Hackett, M. Shoup, and S. Slattenow. 2000. *The Role of the Pipe-Water Interface in the Formation of Disinfection By-Products and the Loss of Disinfectants in Drinking Water*. Denver. Colo.: AwwaRF and AWWA.

Vandermeyden, C., and D. Hartman. 2001. Water Quality Modeling of a Complete Distribution System: Development, Results and Practical Applications. In *Proc.* of the *AWWA Annual Conference*, Washington, D.C.

Vasconcelos, J.J., W.M. Grayman, L. Kiene, O. Wable, P. Biswas, A. Bhari, L.A. Rossman, and R.M. Clark. 1996. *Characterization and Modeling of Chlorine Decay in Distribution Systems*. Denver, Colo.: AwwaRF and AWWA.

Vasconcelos, J.J., L.A. Rossman, W.M. Grayman, P.F. Boulos, and R.M. Clark. 1997. Kinetics of Chlorine Decay. *Jour. AWWA*, 89(7):54-65.

Vikesland, P.J., K. Ozekin, and R.L.Valentine. 2001. Monochloramine Decay in Model and Distribution System Waters. *Water Research*, 35(7):1766-1776.

Vikesland, P.J., and R.L. Valentine. 2000. Reaction Pathways Iinvolved in the Reduction of Monochloramine by Ferrous Iron. *Environ. Sci. and Technol.*, 34(1):83-90.

Volk, C. E., J. Dundoree, M. Schiermann, and M.W. LeChevallier. 2000. Practical Evaluation of Iron Corrosion Control in a Drinking Water Distribution System. *Water Research*, 34(6):1967-1974.

Weber, W.J., and F.A. DiGiano. 1996. *Process Dynamics in Environmental Systems.* New York: John Wiley & Sons.

ABBREVIATIONS

Å	angstrom
α	roughness constant
°C	degrees centigrade
C	free chlorine concentration
C_A	chlorine concentration at Point A
C_b	concentration of disinfectant in the bulk
C_B	chlorine concentration at Point B
CF	C-factor in Hazen-Williams equation
CI	cast iron
cm	centimeter
CMBR	completely-mixed batch reactor
CMFR	completely-mixed flow reactor
C_o	initial free chlorine concentration
CR	corrosion rate
C_t	free chlorine concentration at any time
C_w	concentration of disinfectant at the wall
$\Delta C_{t,b}$	chlorine$\Delta C_{t,w+b}$ = chlorine consumed after time, *t*, in the pipe-section reactor due to both wall and bulk decay
$\Delta C_{t,w+b}$	chlorine consumed after time, *t*, in the pipe-section reactor due to both wall and bulk decay
d	day
D	pipe diameter
DI	ductile iron
DO	dissolved oxygen
DS	distribution system
F	fractional decrease in chlorine attributed to wall reaction
ft	feet
ft/d	feet per day
g	gram
g	acceleration of gravity
h	head loss
hr	hour
k	first-order rate coefficient
K	experimentally determined rate coefficient
k_b	first-order rate coefficient for bulk chlorine decay

$k_{b,field}$	first-order bulk decay coefficient calculated at field temperature
$k_{b,lab}$	first-order bulk decay coefficient calculated at laboratory temperature
k_f	mass transfer coefficient
km	kilometer
k_{ps}	pseudo first-order decay coefficient
k_{tot}	sum of k_b and k'_w
k_w	first-order rate coefficient for wall reaction
k'_w	first-order rate constant in units of time^{-1}
K_w	zero-order rate coefficient
k_{w+b}	rate coefficient from DI pipe reactor test
L	liters
L	pipe length
m	meter
M_{Cl}	atomic weight of chlorine
m^3/d	cubic meters per day
M_{Fe}	atomic weight of iron
mg	milligram
mgd	million gallons per day
mm	millimeter
mpy	mils per year
m/s	meters per second
μA	microamp
μm	micrometer
μohms/cm	microohms per cm
N	flux of chlorine to pipe wall
ntu	nephelometric turbidity unit
OWASA	Orange Water and Sewer Authority
PLR	propeller loop reactor
$PS_{3''\text{-}D}$	power setting of the rheostat to control the stirrer motor for 3-inch diameter Propeller
$PS_{4''\text{-}D}$	power setting of the rheostat to control the stirrer motor for 4-inch diameter Propeller
PVC	polyvinyl chloride
r	pipe radius
R^2	correlation coefficient
r_h	hydraulic radius
ρ_{Fe}	density of iron
s	second

t	time
$\mathrm{T}_{\mathrm{field}}$	field temperature
T_{lab}	laboratory temperature
TOC	total organic carbon
v	water velocity
v_{ball}	velocity of polyurethane foam ball
V_{cor}	corrosion velocity
v_{Pitot}	velocity measured by Pitot tube
WTP	water treatment plant

IWA Publishing's authorised EU representative for General Product Safety Regulations is Diane D'Arras, 15 rue Duret, 75116 Paris, France, e-mail: safety@iwap.co.uk.

Printed and bound by CPI Group (UK) Ltd, Croydon, CR0 4YY
12/05/2026
02108865-0009